A MOST IMPROBABLE JOURNEY

A MOST IMPROBABLE JOURNEY

A BIG HISTORY OF OUR PLANET AND OURSELVES

WALTER ALVAREZ

W. W. NORTON & COMPANY

INDEPENDENT PUBLISHERS SINCE 1923

NEW YORK • LONDON

For information about permission to reproduce selections from this book,
write to Permissions, W. W. Norton & Company, Inc.,
500 Fifth Avenue, New York, NY 10110

For information about special discounts for bulk purchases, please contact
W. W. Norton Special Sales at specialsales@wwnorton.com or 800-233-4830

Manufacturing by Quad Graphics Fairfield
Book design by Brooke Koven
Production manager: Julia Druskin

ISBN 978-0-393-29269-5

W. W. Norton & Company, Inc.
500 Fifth Avenue, New York, N.Y. 10110
www.wwnorton.com

W. W. Norton & Company Ltd.
15 Carlisle Street, London W1D 3BS

1 2 3 4 5 6 7 8 9 0

A Most Improbable Journey is dedicated
to the memory of Jack Repcheck—
a dear friend, the best publisher an author could hope for,
a fine author himself, a historian, a musician, and
a good and decent man who lived his life well.
This is Jack's book, too, though tragically
he did not live to see it finished.
Requiescat in pace.

Contents

PROLOGUE

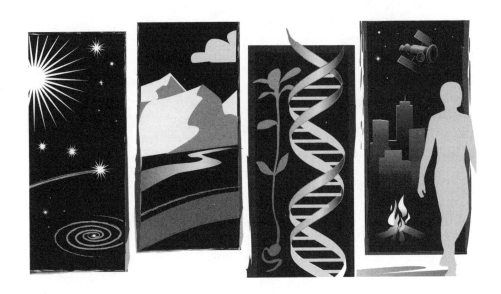

INTRODUCTION

THINK FOR A moment about the human situation we live in—the solar system and our planet; the continents, oceans, and landscapes; the animals and plants; the nations, governments, and businesses; all the languages, cultures, and beliefs; our particular city or town; our family and all the people we know. How did all this come to be? To understand the human situation we need to consider the history that has produced it.

However, for those of us who are fascinated by history, most of the books we read are quite specialized—their subjects are restricted in time and space. We may find books about the French Revolution, the American Civil War, the Ming Dynasty in China, or the Spanish discovery and colonization of Latin America. From focused books like these, it is hard to see how everything fits together. To counter this kind of specialization, there are also books dealing with World History—unifying all of the human past on a global basis.

But to a historical *scientist* like me—a geologist—or to a paleontologist or an astronomer or an archaeologist, even a book about World History is narrowly restricted in time and space! Right now historical scientists are delighting in the discovery of an enormous history, reaching back billions of years into the past, and stretching across a vast universe of which we are simply one local neighborhood. Our human history is only part of the story of the past, although a fascinating part indeed.

The broader history of everything might seem irrelevant to someone interested in human history, but it is not. The human situation in which we find ourselves is the result of a history that has unfolded across enormous stretches of time and space, and almost everything that has taken place in human history has been strongly influenced by events deeper in the past.

Those of us who are interested in understanding the entire past now call this panoramic viewpoint "Big History." I like to think of Big History as combining four regimes—Cosmos, Earth, Life, and Humanity. Each of the four regimes is filled with fascinating stories that help us understand what it is to be a human being, living in our particular world, and not some other kind of creature living in some other place.

For me, the astonishing realization that comes from the study of Big History is just how unlikely our world is. At innumerable points in its history, events could have led to totally different results—to a human situation completely different from what we know today or to a world with no humans at all. Our history has been *A Most Improbable Journey*, and that will be the through-going theme of this book.

The first three regimes of Big History are the subject of scientific research, not humanistic scholarship, and so they may be less familiar to those who read history for pleasure. My goal in this book is to make all that prehuman history accessible and easily comprehensible, no matter what your background is in humanities or science. Although there is a continuity from chapter to chapter, feel free to read them in any order that interests you. If *A Most Improbable Journey* leaves you with a greatly expanded view of what history is about, with a new appreciation of the human situation, with a delight in a whole series of fascinating stories, and with a lot of new questions, then it will have been a success.

Welcome to Big History!

CHAPTER ONE

BIG HISTORY, THE EARTH, AND THE HUMAN SITUATION

AN EXPEDITION TO MEXICO

I T DID NOT start out like an unforgettable day—that Tuesday in February of 1991 in the lowlands of eastern Mexico. It started out with two broken-down jeeps and a whole load of trouble getting even one of them fixed. By the time we were able to leave Ciudad Victoria for the field, it was afternoon and a good bit of our last day was already gone. Jan, Sandro, and I, along with a post-doc named Nicola, were trying to find ancient debris ejected from the recently discovered Chicxulub Crater in the Yucatán Peninsula, several hundred miles away. We believed that the crater had been formed by a huge object from space striking the surface—probably an asteroid but possibly a comet. We had already spent three days in a fruitless search for the ejecta all across northeastern Mexico, and it looked like one final day of frustration awaited us.

About 12 years earlier, the young Dutch geologist, Jan Smit, and I had independently discovered unexpectedly high concentrations of the element iridium in a clay layer that separates sedimentary rocks of Cretaceous and Tertiary ages.[1] Unaware of each other, we had each been trying to understand the reason for the great mass extinction of life that occurred 66 million years ago, at the end of the Cretaceous. That mass extinction had spelled the end of the dinosaurs. Iridium is extremely rare in Earth's surface rocks but somewhat more abundant in comets and asteroids. So, in studies with our colleagues, we both proposed that the iridium must have come from an extraterrestrial source. We further proposed that Earth had been hit by a very large comet or asteroid on that very fateful day 66 million years ago.[2]

Jan and I had become close colleagues and friends in the decade or so of intense scientific argument over whether the impact hypothesis was correct. Even though evidence kept piling up in support of an impact 66 million years ago, skeptics demanded to know where the crater was located. If our impact hypothesis was correct, there had to be a crater somewhere. But no one could find it.

One line of evidence was particularly compelling. Sandro Montanari, a young Italian geologist who did his PhD with me, discovered tiny round objects—he called them spherules—in the sedimentary layer that marks the Cretaceous-Tertiary boundary in Italy. Jan also found them in Spain, so together we published an interpretation that has held up ever since: These spherules were formed from droplets of rock that had been melted by the heat of the impact, were ejected from the impact crater, left Earth's atmosphere and traveled long distances in ballistic free fall before reentering the atmosphere and falling back to Earth.[3] The spherules provided very strong evidence for a huge impact but still did not answer the question of where the crater was located.

Now, at last, there was a candidate crater, and we had come to Mexico to put it to the test.

MIMBRAL QUEST

Something like half of all the genera of plants and animals living on Earth 66 million years ago perished in the great mass extinction that geologists use to mark the Cretaceous-Tertiary boundary—the most recent of six mass extinction events that have punctuated life history. Jan, Sandro, and I, as well as a number of other geologists studying the Cretaceous-Tertiary boundary, were proposing that the extinction was caused by a huge impact, and most geologists and paleontologists did not like it at all. Our idea went completely against their "uniformitarian" training.

What was "uniformitarian" training? Charles Lyell, an early English geologist, had argued forcefully in the 1830s that all changes in Earth's past have been slow and gradual. This doctrine is called Uniformitarianism, and by the twentieth century, it was treasured and defended by geologists everywhere. Our theory was a direct threat to Uniformitarianism. Nicola Swinburne, the post-doc who was with us on that February day, had done her PhD in England where geologists were particularly sensitive to threats to Uniformitarianism. She would remain skeptical of the impact hypothesis, despite the amazing evidence we were about to find.

About 50 miles south of Ciudad Victoria, we turned left off the highway and found a rough gravel road along a nearly dry valley called Arroyo el Mimbral. This was our final target. A few months earlier in the library at Berkeley, I had found a 1936 book about the geology of this region that described a strange sand bed exposed along the Arroyo el Mimbral.[4] The sand bed occurred within deep-water clays that had been deposited on the floor of the Gulf of Mexico in the late Cretaceous and early Tertiary and had subsequently been raised up and exposed on land. Could this sand bed be ejecta from the Chicxulub Crater? We surely hoped so!

If we found an impact ejecta bed exactly at the Cretaceous-Tertiary boundary, it would be compelling evidence that the nearby but deeply buried Chicxulub Crater was exactly the same age as the mass extinction and that it was the missing crater we had been seeking for a dozen years. It would provide strong evidence in support of the impact hypothesis.

At the surface of the Yucatán Peninsula you can see nothing unusual, but Mexican petroleum geologists had found the buried crater many years earlier by detecting slight variations in the pull of gravity. In 1981 the Mexican geologist Antonio Camargo-Zanoguera and his American colleague Glen Penfield had proven beyond a doubt that these variations were caused by a great impact crater, 120 miles in diameter, buried by about a mile of younger sediments. However, their company, Pemex, had not allowed them to publish their findings. Finally, in 1991, a Canadian graduate student, Alan Hildebrand, heard about their results, and together they were finally allowed to publish the impact interpretation of Chicxulub.[5]

Their paper was a bombshell! The largest impact crater known on Earth—except for two extremely ancient ones—was at least approximately the same age as the mass extinction. Was it exactly the same age? Was it the crater we had been seeking for a dozen years? As we bounced along the rough road following the Arroyo el Mimbral, worrying as the Sun got lower in the sky, we hoped that there would be impact ejecta at exactly the extinction level, but we could not have imagined what awaited us.

MIMBRAL DISCOVERY

Not every geologist knows how to find the Cretaceous-Tertiary boundary. But Jan does. Using a little magnifying hand lens, he can identify the tiny microfossils that tell the age of the sediment. Every now and then, we would stop at some little outcrop of clay, and Jan would look at it carefully. "We're coming down through the lowermost Tertiary, getting close to the boundary," he would announce. The beds of clay sloped, or dipped, very gently toward the west, away from the mountainous Sierra de Tamaulipas, so we had to drive a long way to get very much farther down in the sequence of clay layers. There were few hills and almost no outcrops, as you would expect in a landscape of soft clay, and we were getting discouraged.

And then we saw it. On the other side of the arroyo, the dry riverbed, there was a steep bluff with spectacular rock exposures. Dense brush blocked the way, but a quarter mile or so farther the road crossed the arroyo, and we almost ran back along the stream bottom to get to the bluff. It was the most wonderful outcrop I have seen in five decades as a geologist.

Fig. 1-1. Jan Smit at Arroyo el Mimbral, on the evening we discovered this amazing outcrop.

You could tell immediately that something dramatic had happened here. At the bottom of the cliff, Jan found microfossils from near the very end of the Cretaceous. At the top of the cliff, there were microfossils from the very oldest Tertiary. And in between was the massive sand layer mentioned in the 1936 book. If it was ejecta from Chicxulub, then the crater was exactly the right age!

Sand is one of Nature's most common sediments—geologists see it all the time—but this was like no sand any of us had ever seen before. We each scrambled up different parts of the bluff, trying to see as much as possible in the fading light, and shouting out what we were finding.

The fine-grained clays above and below the sand bed could only have been deposited in very still water, deep in the Gulf of Mexico, and there should never be waves or strong currents in water that deep. The water should have been quiet and calm, barely moving at all, but the sand was full of little layers sloping in different directions. Geologists call these sloping layers "cross-beds," and they show that rapid currents at great depth were involved. Near the bottom of the sand layer were chunks of the underlying clay that had been ripped up and carried along by violent currents. Something bad had happened on the floor of the Gulf of Mexico!

Nicola yelled for us to come and see what she had found—a bed of fine sand made just of microfossils, but scattered through it were spherules the size of peppercorns. Calling it the Nicola bed, we guessed that the spherules had been droplets of impact melt, and laboratory work later on confirmed that guess. Even in the growing darkness we could see tiny bubbles in the spherules and guessed, correctly, that they were due to CO_2 gas released from Yucatán limestone by the energy of the impact. Limestone is made of calcium carbonate—$CaCO_3$—and it gives off CO_2 when it is heated, so this was direct evidence for impact at the Cretaceous-Tertiary boundary. The very base of the sand bed was packed with bigger spherules full of gas bubbles and with pieces of Yucatán limestone, never melted, that had traveled outside the atmosphere for several hundred miles as a result of the impact.

Sandro happened to look up at the base of an overhanging sand bed and saw that it was felted with pieces of wood, now petrified. You do not find wood on the deep-sea floor, certainly not great masses of it. We were mystified at the time, but later we understood its significance. Careful study has shown that the spherules arrived on the fly just as a great tsunami from the impact site reached the Mimbral area, tearing up the sea bottom and continuing on, demolishing what was then the coastline of Mexico. The coastal sediments, saturated with water, flowed back down into the deep Gulf of Mexico. The felted wood came from the coastal forests, destroyed by the tsunami. At the top of the sand were beds with smaller-scale cross-beds, suggesting that the remains of the tsunami had sloshed around several times in the Gulf before finally calming down.

It was an amazing story recorded in a most amazing outcrop. It was the kind of discovery scientists dream of but rarely experience.

BIG HISTORY

The Mimbral outcrop was obviously so important that Sandro and Jan changed their plans, camped there for several days, and came to understand in depth the story the rocks were telling. Back in Berkeley, we sent samples to colleagues with specialized laboratories, and in 1992 we published a major paper about the Mimbral discovery.[6] The next year we made a second trip to Mexico, with Mexican geologists, discovering several more remarkable Cretaceous-Tertiary outcrops, each one telling more about the nature of the impact-extinction event.[7]

Since then we have traveled somewhat different paths. Sandro returned to Italy and founded the Geological Observatory of Coldigioco as a base for reading all kinds of Earth history recorded in the wonderful deep-water limestones of the Apennine Mountains. Jan has continued in his fascination with the Cretaceous-Tertiary boundary, visiting and studying more boundary outcrops all over the world than any other geologist, extracting more and more information about the detailed history of that event. If Jan has studied the boundary event through a microscope, as it were, then I've looked at it through a telescope, backing off farther and farther in order to understand impact and mass extinction in the broadest possible historical context.

I've always been fascinated by history of all kinds. As a geologist my main expertise is in Earth history, but I've also had the opportunity to learn about life history because of the mass extinction, and about cosmic history because of the great impact of the asteroid or comet. In addition, I've always been interested in human history because of the unusual places to which geology has taken my wife Milly and me, but for a long time I thought of that as no more than a hobby.

Eventually I started wondering if it might be possible to combine all those different kinds of history into some sort of interdisciplinary field with an overarching view of all of the past. And then one day, I received a letter from the Dutch biochemist and anthropologist Fred Spier, telling me about "Big History," which aims to do just that.

The concept and the name are due to the Australian historian David Christian, who was attempting to get away from the extreme specialization of most historians.

So, with a Berkeley grad student named David Shimabukuro, who has unusually broad interests, we developed a course in Big History at Berkeley, and it was the most intellectually exciting teaching experience I've ever had. David and I found that although many Berkeley students were happy to become narrow specialists, there were others who were just craving to see how all their specialized courses fit together. I have never seen students as excited to take a course as they were with Big History. One of those students, Roland Saekow, suggested that we develop a zoomable, computer-graphics time line of all of history, back to the Big Bang. Roland ended up working with David and me and with Microsoft Research to develop ChronoZoom, available now online, at ChronoZoom.com.

For organizing our course, we found it useful to think of Big History as divided into four regimes—Cosmos, Earth, Life, and Humanity. Big Historians have been finding out that the academic chasm between the sciences that study the first three regimes and the humanities and social sciences that study the fourth is truly difficult to bridge but that the challenge brings great rewards.

The human situation

In our course we tried to increase our understanding of the situation in which we humans find ourselves by developing "historical mindedness"—the habit of thinking historically about whatever we encounter in our lives, reaching across the whole range of Big History, from the beginning of the universe to today. We found that historical mindedness offers wonderful insights into the human situation.

To begin with, physics and chemistry lie behind everything we encounter in all of the human situation. The great discoveries of physics—orbital motion, electromagnetism, relativity, quantum mechanics, and chemical bonding—have a great deal to say about how the world works and about the natural laws that control its behavior. But they tell us little about how our *particular* world came to be. Why this world, and not some other, equally plausible world that could have arisen with the same laws of physics and chemistry?

As we will see in the last chapter, history is contingent—chance has played an enormous role in history. At innumerable moments throughout the regimes of Cosmos, Earth, Life, and Humanity, history could have taken different paths than the one our world actually did take, resulting in a human situation different from the one we have today—or possibly no human situation at all!

To understand our particular world we need to go beyond physics and chemistry, into the realm of the *historical* sciences—geology, paleontology, biology, archaeology, astronomy, and cosmology—and then on to human history. We need to know what historical scientists and historians are learning about the particular history that actually did take place. To appreciate the distinction between the physicists' world and that of the geologists, archaeologists, and astronomers, we can think about a mildly disturbing painting by the Belgian artist, René Magritte.[8]

Magritte painted a large rock, surmounted by a castle, placidly floating in the air above a tranquil seascape. Rocks are part of the human situation, but the image disturbs us because we know that rocks do *not* float in air—this is *not* part of the human situation. A rock may find itself in this position for the briefest instant, but only if it is an asteroid, hurtling toward Earth at 70,000 miles per hour, accelerating as gravity pulls it down, and heated to incandescence by friction with the air, an instant before it crashes into Earth, excavating a huge impact crater. This behavior of real rocks is the kind of thing we can calculate on the basis of the mathematical laws of physics.

Geologists make use of the physical laws of falling rocks but are more interested in specific instances of specific falling rocks that have been important in history. Our discovery of the Mimbral outcrop in 1991 was important because it demonstrated that a particular falling rock, 66 million years ago, coincided exactly in time with a particular event in life history—a great mass extinction. The mass extinction was critical in creating the human situation—quite literally—because without the impact and the resulting extinction, dinosaurs would in all probability still be the largest animals on Earth, mammals would still be small, and there would be no humans even to *have* a situation. Here is a first, and very dramatic, example of the improbable historical journey that has led to us.

History lies behind every part of the human situation, but the usual, specialized kind of history doesn't really help us understand our whole situation. *Big* History is the tool we need.

THINKING ABOUT BIG HISTORY
AND THE HUMAN SITUATION

Anywhere we look, we can see examples of the almost infinite aspects of the human situation—from the broadest in a map or satellite view to the most detailed in a close-up photo, a book, an organizational chart, or with a microscope. For example, in this satellite image of North America at night, think about all the aspects of the human situation it suggests and all the history that lies behind it—from the East Coast where Africa broke away 180 million years ago, to the distribution of population reflecting the westward expansion and the availability of water (far fewer towns and cities in the West), to the differences between Mexico, the United States, and Canada. And think about all the institutions of the human situation that lie submerged in the pattern of lights—government, industry and commerce, science and high technology, universities and colleges, military bases, railroad lines and roads, religious groups, and millions of families and individuals, each with their own histories and peculiarities.

So how does a Big Historian begin to think about the human situation? Questions are everywhere, and one approach is to try to formulate broader questions than scientists and scholars normally consider.

An astronomer would use the laws of gravity and of orbital mechanics to calculate the trajectory that leads an asteroid to crash into Earth, but a Big Historian would like to know how and when

Fig. 1-2. A broad overview of part of the human situation, portrayed in a satellite view of North America at night.

gravity itself came into being. Has there always been gravity, or did it emerge at a specific time?

Geologists tend to focus on details of geological history, such as the origin of a particular mountain range, but a geologist guided by Big History might want to understand the whole sweep of continental motions throughout all of Earth history that has given rise to all the mountain ranges. Have there been recognizable patterns, regularities, cycles, and contingencies in the history of continental motions?

Many biologists spend their time trying to understand the complexities of particular animals and plants, but a Big Historian might try to understand why organisms are complex at all, and whether complexity has developed and changed through time. Have new kinds of complexity appeared at particular historical moments?

Most historians of humanity are interested in particular contingent events that have led to the particular human situation we live in today, but a Big Historian might want to understand the nature of contingency, as we will try to do in the final chapter.

A second approach is to try to understand all the history that lies behind some particular feature of the human situation. The Dutch Big Historian Esther Quaedackers pioneered this approach, and she calls this kind of study a "little Big History." Almost anything could be the subject of a little Big History. A drinking glass, perhaps. Glasses come in all kinds—round or square, tall and thin or short and fat, with and without stems, plain or ornamented, clear or colored. When, where, and why did each kind of drinking glass appear? When and where did people learn to make glass? How did Earth concentrate the quartz sand that is the raw material for making glass? Quartz is made of silicon and oxygen, but how and when did they appear in a universe that started out as just hydrogen and helium?

A more challenging little Big History might try to convey an understanding of a mountain range like the Alps. How have the Alps affected human history? Would human history have been different if there were only flat land between Italy and Germany? How and when did the particular topography of the Alps—the Matterhorn, for example—come to be? Why is there a mountain range there at all—can it be related to the history of continental motions? How and when did each complicated body of Alpine rocks come to be? What particular history brought to Earth the particular elements that are needed to form those rocks?

Or we could do a little Big History of a great language like Spanish, thinking about how this particular descendant of Latin came to dominate the Iberian Peninsula and then much of Latin America. Were there geographic controls and, if so, what geological history lay behind them? How and when did language itself appear, and what anatomical features gave humans alone the capability for complex language? How has the dominance of this particular language affected the unrolling of human history in the last millennium?

In this book there will be examples of both broad questions and little Big Histories. My background is in geology—I'm a historian of the Earth. So rather than trying to portray the history of the human situation in a way that is both very broad and very detailed, I've chosen to look at the condition of humanity from a geologist's point of view. People like me always think first about our Earth—its deep interior, its surface features, its oceans, and atmosphere. This is not common among others who try to understand Big History or the human situation, but perhaps it will be refreshing—providing a new way of thinking about our world.

So let us begin with a quick account of how cosmic history has produced the planet and the solar system where we live. Then we will focus more carefully on Earth and on the great variety of plants and animals that ride on it. And finally we will think about some fundamental characteristics of humanity and about how our Earth has conditioned their development.

COSMOS

CHAPTER TWO

FROM THE BIG BANG
TO PLANET EARTH

THE MEANING OF "AWESOME"

THE HUMAN SITUATION begins with the fact that we humans live in a Cosmos that is absolutely enormous. We can measure it, but we have no words that convey its scale in a way we can really understand. The Sun, which dominates our sky and gives us life, is an ordinary, run-of-the-mill star. This star of ours is just one among a vast, vast throng of stars making up a single galaxy that we call the Milky Way. And our Milky Way itself is an ordinary, undistinguished galaxy among a vast, vast throng of galaxies stretching in every direction as far as our most powerful telescopes can see.

All the history that has produced our global civilization has played out on this planet, which is an utterly negligible speck in its cosmic context.[1] We must begin our exploration of Big History with this overwhelming lesson in humility, but we will then discover that within this vanishingly tiny realm of Earth, the history that has led to us is itself endlessly fascinating.

We can write down the numbers and say that our galaxy, the Milky Way, includes something like 100,000,000,000 stars (a hundred billion), and that it is one of about 100,000,000,000 galaxies (also a hundred billion). Scientists prefer the exponential notation, counting the number of zeros and saying that there are roughly 10^{11} galaxies, each

Fig. 2-1. The awesome scale of the Cosmos, zooming out. *Top*: The Sagittarius Star Cloud—a dense cluster of stars within our own Milky Way Galaxy. *Middle*: The Southern Pinwheel Galaxy—a single galaxy, probably much like ours. *Bottom*: The Hubble Ultra Deep Field—every bright spot in this tiny fraction of the sky is a very distant galaxy.

with about 10^{11} stars, or about 10^{22} stars in the Cosmos. The exponential notation is more compact and easier for doing calculations, but it obscures the horrifying fact that our Sun is just one of about 10,000,000,000,000,000,000,000 stars altogether.

We commonly hear people say that a particular musician or some basketball player is "awesome." What a foolish misuse of a powerful word! If you want awesome, try thinking of a universe containing 10,000,000,000,000,000,000,000 stars and probably at least that many planets.

We almost automatically ask what this truly awesome Cosmos is for—why does it exist? The question is profound, but it is not a question science knows how to answer, and scientists are uncomfortable with it because we have never been able to discern purpose in Nature. But we *can* ask when it came into being and how it has evolved to its present state. We now have preliminary answers to those questions because in the last few decades astronomers and cosmologists have come to the point where we have a basic understanding of the history of the Cosmos.

However, until about 50 years ago, astronomers and cosmologists were engaged in a great debate: Has the Cosmos always existed in roughly its present configuration, in what they called a steady-state universe? Or did it have an actual beginning—an initial moment called the Big Bang? A steady-state universe, which always looks about the same, effectively has no history, but a Big Bang universe is a historical thing. We now know, beyond any doubt, that the Cosmos did begin with the Big Bang, that it *has* a history, and a most fascinating history indeed.

Cosmic history is the first regime in Big History, but here we come to a strange contradiction. It is obviously ludicrous to think that a Cosmos containing this awe-inspiring number of stars was made to be a nursery for the human species, living on one planet, orbiting a run-of-the-mill star in an ordinary galaxy. But the enterprise of Big History is intended to put the history of humanity into context, and from our parochial perspective, all that cosmic history is prologue. We cannot understand the human situation without some knowledge of the cosmic history that prepared the stage upon which human history has played out. So we need to be able to keep two contradictory notions in mind—that from the standpoint of the Cosmos* we are utterly insignificant, while from our own standpoint cosmic history is our heritage.

Let us begin at the beginning, with the discovery of the Big Bang, and I want to tell it in a rather unusual way, focusing on a young man who made his living driving mules.

THE MULE SKINNER'S DISCOVERY

The story of the Big Bang has been told many times, and usually it starts with the astronomer Edwin Hubble. But it seems to me that Hubble has already received plenty of well-deserved recognition, including his name on the great space telescope, so I prefer to begin the story with someone else—with a boy who dropped out of school at age 14 to drive a team of mules and had no further formal education.

His name was Milton Humason (pronounced "Hum," not "Hume") and he was born in Minnesota in 1891, just two years after the birth of Edwin Hubble.[2] Having fallen in love with Mount Wilson while at a summer camp above the little town of Los Angeles, Humason quit school and went to work there as a mule skinner—driving a team of mules in an age before truck transportation became important.

The atmospheric inversion responsible today for the smog of Los Angeles makes the air *above* the inversion particularly stable and ideal for astronomical telescope observations, and the great 100-inch telescope was being installed on Mount Wilson, above the inversion. Completed in 1917, it would be the number-one astronomical instrument for the next 30 years. Milton Humason was hired with his mules to help move lumber for the observatory up to the mountaintop and probably never imagined what would follow.

While moving the lumber, Humason fell in love with Helen Dowd, daughter of the observatory's chief engineer. They were married, and two years later Humason secured a humble job on the mountain as observatory janitor. Fascinated by what the astronomers were doing, he volunteered to help the night assistants develop the photographic plates made by the telescope. He then became a night assistant himself. Humason soon acquired such technical skill that the observatory director, George Ellery Hale, made him a permanent staff member in 1919, an unheard-of success for a high school dropout, but Hale's judgment would soon be vindicated.

In 1915, two years before the 100-inch telescope was finished, Einstein published his general theory of relativity. Einstein's concept that massive bodies like planets, stars, and galaxies distort the geometry of spacetime is still the basis for understanding gravity, but not in exactly the form that he proposed. Although today it seems puzzling, Einstein was convinced that the universe is static and ahistorical— neither expanding nor contracting with the stars permanently fixed in position.[3]

As the most powerful astronomical instrument anywhere, capable of seeing farther than any other telescope, the 100-inch telescope on Mount Wilson was the place to test whether the universe is static. Edwin Hubble was the man for the job, but by all accounts he was not a skillful observer and was not able to make the necessary measurements himself. Fortunately he had Milton Humason to work with.

What Humason and Hubble found would radically change our understanding of the universe. Night after night, Humason would point the great telescope at galaxies beyond our own Milky Way Galaxy, measuring how fast they are moving toward or away from us (using the "red shift" of absorption lines in the spectrum of their light) and estimating their distance away from us, which was very difficult to measure accurately. The spectral lines showed that all but the closest galaxies are moving away from us, and the farther away they are, the faster they are receding. This sounds like we are at the center of everything, but in fact every astronomer in every galaxy will see the same thing if the universe is expanding, and that is indeed what is going on.

The discovery is called Hubble expansion, and the ratio of galaxy recession rate to distance away is called the Hubble constant, because

Fig. 2-2. Milton Humason, co-discoverer of the expanding universe, at the Mount Wilson Observatory.

Hubble put only his own name on the 1929 discovery paper. But this was clearly a collaboration, for the paper immediately before the 1929 Hubble paper in the Proceedings of the National Academy of Science, which was based on a few nearby galaxies, is by Humason, giving the higher recession rate for a more distant galaxy. It seems to me that astronomers should call it Hubble-Humason expansion and the Hubble-Humason constant.

In an expanding universe, if you were to go backward in time, the galaxies would get closer and closer together, until all the galaxies and all the space between them would be confined to a tiny ball, and this was the Big Bang, almost 14 billion years ago. The Big Bang is usually described as an explosion, although not like ones familiar to us. It was not an explosion *within* space, like a firecracker or a quarry blast, but an explosion *of* space and of matter and even of time itself, none of which existed until the explosion took place. This is the physicists' and cosmologists' current picture of the beginning of everything, including time, although this explanation is very hard to grasp intuitively. But as Harvard cosmologist Lisa Randall once told me, we really don't *know* what happened at the very beginning.

The expanding universe was unquestionably one of the greatest scientific discoveries, so now it is time to stop thinking of Milton Humason as the dropout mule skinner and janitor of his early years and remember him instead as the truly great scientist he became. And in fact, in 1950 he became *Doctor* Humason, when he received an honorary doctorate from the University of Lund in Sweden. An ordinary PhD rewards a promising piece of initial research by a young scientist. An honorary doctorate recognizes a lifetime of great work by a mature scientist. Perhaps never was one more deserved than the doctorate received by Milton Humason, because it was his work with Hubble that, after a long controversy, eventually established that the universe is not eternal—it is of finite age, and it *has* a history.

GEOLOGY AND THE BIG BANG

What could the history of an insignificant little dot like Earth have to tell us about the Big Bang and the history of the entire Cosmos? Quite a lot, as it turns out. . . .

The discovery of cosmic expansion is usually told more or less as I have just told it, often with more detail and generally with Hubble as the solitary hero, but the story is actually much more complex and interesting. Astronomers did not easily accept expansion and the corollary that came to be called the Big Bang, and the reason is typical of the fits and starts, the mistakes and corrections of the scientific endeavor.

It was extremely difficult for Humason and Hubble to measure or even estimate the distances to remote galaxies. Only the nearest stars can have their distances determined directly, by measuring their parallax—how much they move against the distant background stars as Earth circles around the Sun. For stars farther away and for galaxies, astronomers measure the apparent brightness of objects whose *true* brightness they think they know—the quaintly named "standard candles"—just as you could estimate the distance to a town at night by the apparent brightness of its streetlights. The scale of the

Cosmos is so enormous that astronomers have found a whole series of standard candles that work at different distances, and they string them together to make a "cosmic distance ladder" with uncertainties on every rung, uncertainties that build up so that the distances to remote galaxies were extremely uncertain in the time of Humason and Hubble in the late 1920s.

We now know that their estimates for distances to remote galaxies were very seriously in error—they underestimated the distances by a factor of about seven!—and this had grave consequences for understanding cosmic history, and whether the Cosmos even *has* a history. Knowing the velocities at which each galaxy was receding, and thinking they knew its distance, they could project back to a time when all the galaxies were clustered tightly together, which we now call the Big Bang.

Here was the problem:[4] Using inaccurate distances, they miscalculated that the time since the beginning was only 2 billion years, which proved to be too brief. By the 1930s, geologists were learning to date the age of Earth by radioactive decay in minerals and determined an age of 1.6 to 3.0 billion years old.[5] This was barely enough time, or not enough time at all, for the history of Earth to fit inside the history of the Cosmos!

Something was seriously wrong, and it is clear that the astronomers took the geologists' age of Earth very seriously. The problem of having an Earth that is older than the universe was so critical that Edwin Hubble, now canonized as the discoverer of the expanding universe, stopped believing his own expansion theory within a few years! From 1935 until his death in 1953, Hubble argued that the red shift of distant galaxies was not due to their velocity away from us but to some other phenomenon that had not yet been understood. What an irony, that its very discoverer came to reject the fundamental basis of modern astronomy and cosmology![6]

In response to this dilemma, skeptical cosmologists invented the steady-state theory as an alternative to the Big Bang. They proposed that the universe does indeed expand, but that new matter is continually created to fill in the space left by expansion, so that the age of the universe could be very great or even infinite. The steady-state theory is just a curiosity today because now we know why there was a problem. The solution to the problem was much less exotic than continual creation of new matter. It was simply that early distance measurements were wrong. By the 1960s, better telescopes and a better cosmic distance ladder showed that the remote galaxies are about seven times farther away than Humason and Hubble had thought. This corresponds to an age of the universe of 10 billion years or more, so the Cosmos is comfortably older than Earth.

In addition, there is wonderful new evidence confirming the Big Bang—from the cosmic background radiation, a radio-wave buzz emitted when the universe was only about 380,000 years old, as well as the cosmic ratio of the two lightest elements, hydrogen and helium, which is exactly what the physics of the Big Bang predicts. And in fact, the best current age of the universe, from the European Planck satellite, is 13.8 billion years, which allows plenty of time for an Earth now known to be about 4.5 billion years old.

Despite their understandable mistake in the distances, the work of Hubble and Humason was an enormous step forward, laying the foundation for the Big Bang theory and for everything we now know about the history of the Cosmos.[7] Just as biologists say that nothing in the history of life makes sense except in the light of evolution, and geologists understand that nothing in Earth history makes sense except in the light of plate tectonics, for astronomers, nothing in cosmic history makes sense except in the light of the Big Bang theory.

JUST SIX NUMBERS

Although we cannot confidently make observations that take us very far back into the history of the Big Bang, cosmologists can calculate the temperature and the state of matter in those very early times, projecting the universe back toward an initial tiny dot. This book is not the place to explore early cosmic history in detail, but we need to consider a profound mystery implied by the very existence of the Big Bang. This mystery is at the heart of the human situation, lying behind everything in the night sky, every single aspect of the geological planet we live on, the entire realm of living organisms of which we are part, and everything about us as a species of intelligent, communicating, toolmaking primates.

None of those aspects of the human situation could be what they are if the laws of physics, the kinds of matter, or certain fundamental constants were different. Were any of those parameters even a little different from what they are, the Cosmos would very likely be completely different or might not exist at all. Our Sun has burned steadily for a long enough time for life to evolve because those parameters make possible nuclear fusion—the slow release of huge amounts of energy when lighter atomic nuclei combine, or fuse, into heavier ones. The first step is the fusing of hydrogen nuclei to make helium, and the process continues to make heavier and heavier nuclei, all the way out to iron.

Our planet and all living organisms, including humans, are what they are because of the elements that exist and the way they form chemical bonds to make minerals, rocks, and biologically active molecules and because the heat released by fusion in our Sun warms our planet.

So how and when did those physical laws, kinds of matter, and fundamental constants come into being? Light reaching us from galaxies so distant that we are seeing them shortly after the Big Bang behaves just like light from nearby stars, so clearly the laws, kinds of matter, and constants were established by the end of the Big Bang. And cosmologists can calculate reasonable scenarios back to very early in the Big Bang using those parameters, so it appears that they were already in place at or near the very beginning. How were those fundamental parameters locked into the structure of the Cosmos, so that every electron everywhere behaves in just the same way? Nobody knows.

And how do they happen to have just the values they do? The English astronomer Martin Rees has written a profoundly important little book about this part of the human situation, called *Just*

Six Numbers.[8] Rees explored six of the fundamental constants that have somehow been built into the fabric of the Cosmos since the very beginning. He pointed out that current physics has no way to account for the values they have and argued that if any one of them had been different by even a little, the Cosmos also would have been extremely different and would have held no place for us. Let us look briefly at just one of Rees's six numbers.

In the first chapter we considered gravity as a fundamental part of the human situation, explaining why asteroids fall from the sky at enormous velocity, blast open great craters, and do not float placidly above seascapes. The rocks that make up asteroids are held together by electrical forces. Rees points out that the ratio of the strength of the electrical force to that of gravity is 10^{36}, which we can write out as 1,000,000,000,000,000,000,000,000,000,000,000,000, to get a sense of how enormously stronger the electrical force is than gravity. This seems very strange, because we feel gravity all the time and rarely notice electricity. The reason is that positive and negative electrical forces cancel out at short distances, but gravity is always attractive and operates over the scale of the universe.

But suppose this ratio were not quite so large, meaning that gravity was stronger than we experience. In that case stars could be much smaller, more numerous, and closer together; planetary systems like ours would not be stable, and stars would burn out quickly, so there would not be enough time for life to evolve. Or a weaker gravity might have led to a Cosmos *more* complex than the one we live in and, thus, very different. Reading Rees's discussion of the effect of small changes in any one of his six numbers gives a strong sense that the human situation is balanced on a knife edge of improbability, which again is the principal theme of this book.

THE DARK AGE AND THE STARRY EPOCH

In retrospect it seems like the Cosmos was on a dead-end track. It had started, mysteriously, as an infinitesimally tiny ball at enormous temperature—the Big Bang—not *within* space but *including* all space. Expanding very rapidly over the first three minutes after the beginning of time—probably including an episode of hyperfast "inflation"—its temperature cooled down and a whole sequence of subatomic particles flashed into being and transformed to other types in ways that modern particle physics can understand and calculate, and eventually our normal matter of today became stable.[9]

Predominant among that normal matter were protons. Later the Cosmos would cool enough for electrons to combine with the protons to make hydrogen atoms, but before that we can think of the protons as naked hydrogen nuclei. For a brief interval, temperatures were high enough for some of the protons to fuse together and make helium nuclei, but expansion and cooling ended this process when only about 25% of the hydrogen had been converted to helium. So that was the composition of the Cosmos at the end of the Big Bang, after further cooling allowed nuclei and electrons to combine into atoms—about 75% hydrogen, about 25% helium, minute traces of the next element, lithium, and nothing else.[10] It was not a composition that could give rise to rocky planets like Earth, or to life!

As the originally very dense matter of the Big Bang thinned out and became more and more tenuous, it would have appeared that the end of history had come, with nothing left for the future except a more and more dispersed mixture of hydrogen and helium. The Cosmos had entered what astronomers call the Dark Age, because there were not yet any stars. But obviously that was not the end of history, for here we are, on our rocky Earth, bathed in life-giving sunlight! So what happened?

Fortunately for us, Nature had three wonderful tricks that have made our world possible. The first trick was to make stars, the second was to synthesize new elements inside the stars, and the third was to explode some of those stars, releasing the new elements to become part of younger stars and to make rocky planets.

In the first trick, gravity began to slow the expansion of the universe that began with the Big Bang. Cosmic inflation had taken a nearly uniform early universe and magnified its tiny initial fluctuations in density into a universe with major density contrasts. During the Dark Age, gravity pulled the denser regions into ever denser regions that would eventually become galaxies.[11]

Within those protogalaxies were even denser clumps that would become stars as gravity pulled them inexorably to greater and greater density. As those first stars became dense enough for nuclear fusion to take place, they began to shine—to emit light. The Dark Age was over, and the Starry Epoch had begun. It still continues, and today our human situation is a Cosmos with those 10,000,000,000,000,000,000,000 stars.

Around the very earliest stars there could have been gaseous planets like Jupiter, but no rocky planets like Earth because, remember, the matter of the universe was mostly hydrogen, 25% helium, and essentially nothing else. The elements needed to make rocks—mostly magnesium, iron, silicon, and oxygen—did not yet exist.

The second wonderful trick is called nucleosynthesis, because it synthesizes the nuclei of new atoms. The medieval alchemists tried hard to synthesize gold out of common metals like lead, but in retrospect they never had the slightest chance![12] With the little flames in their primitive laboratories, they could induce *chemical* reactions, in which atoms of one element change the binding that attaches them to atoms of other elements, just as we do on the stoves in our kitchens, but the elements themselves do not change.

Chemical reactions involve the loosely bound electrons orbiting at the outer fringe of an atom, but the tiny nucleus sheltering at the center of the electron cloud is utterly impervious to any attack the alchemists could make on it. If the Cosmos had relied on the techniques of the alchemists, we would still have a universe made of nothing but hydrogen.

Yet Nature has laboratories the alchemists could never have conceived of, let alone replicated, and those laboratories are the centers of stars. The real alchemical production of new elements takes place through *nuclear* reactions inside stars. Only in stellar cores are the temperatures and pressures extreme enough to challenge the electrical repulsion that makes protons fly apart or the overwhelming

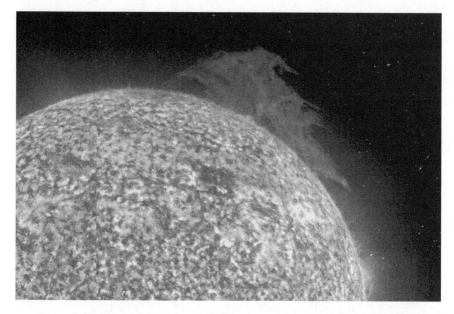

Fig. 2-3. The surface of the Sun, with a gaseous prominence rising to a height 17 times the diameter of Earth. The fearsome temperature at the Sun's surface is merely a pale reflection of the enormously higher temperatures at its core, where nuclear fusion is turning hydrogen into helium.[13]

attraction of the "strong nuclear force" that can even hold protons together.

All the elements from helium, with two protons, up to iron, with 26 protons, are made by fusion, and this includes those four elements that are major constituents of rocks and, therefore, of Earth. The fusion is also critical for us because the heat it releases inside the Sun warms our Earth and makes life possible.

But heavy elements made in stars are trapped inside the stars and are not available to make planets. And what's more, simple fusion does not make any of the elements heavier than iron. Another dead end?

No! Enter the third wonderful trick. As they use up their hydrogen fuel, stars in a certain range of masses will explode. Fortunately our Sun is not in that mass range, but the ones that are will eventually blow up in truly colossal explosions called supernovas, briefly outshining all the other 100,000,000,000 stars in their host galaxy.

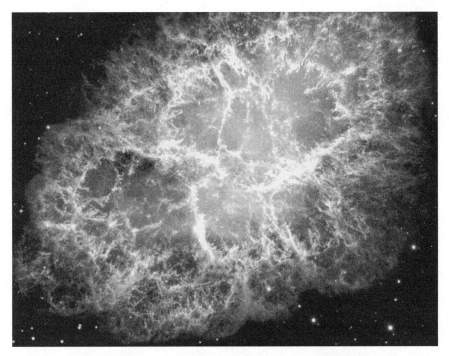

Fig. 2-4. The Crab Nebula, made of filaments of debris dispersing from a supernova explosion observed in AD 1054. The debris is rich in heavy elements made before and during the explosion, which in the future will be incorporated in new stars and planets.

The explosive jamming together of hydrogen nuclei produces all the elements heavier than iron, and the explosion itself scatters every element throughout the region of space surrounding the supernova, where they can be incorporated in new stars. The most recent nearby supernova was observed by Chinese astronomers in AD 1054, and the debris from that great explosion is still visible as an intricate knot of filaments of stellar debris known as the Crab Nebula.

Over billions of years, a long history of successive supernovas, contributing more and more heavy elements to the galaxy, brought it to the point where new generations of solar systems contained enough heavy elements that rocky planets could form. That was the situation about 4.5 billion years ago when our own solar system came into being. Thus it is becoming clear that *matter has evolved* during the history of the Cosmos—an unfamiliar concept but a very important one for Big Historians.[14]

Carl Sagan liked to say that we are made of stardust—from the debris of supernova explosions. This wonderful but nearly incredible scenario is the result of the kinds of particles and the laws of physics that were imprinted on our Cosmos in ways we still do not understand. In the anthropocentric viewpoint of Big History, the stage was now set for the construction of our home planet.

THE BIRTH OF OUR PLANET

This chapter, which opened with a sense of awe at the vast scale of the Cosmos, can now close with a sense of wonder—with amazement that against enormous odds we have this planet Earth, perfectly suited for life and for our own species and its remarkable history.

Our Earth was born amidst almost inconceivable cataclysms. One of the overarching questions of Earth history was memorably phrased in a study of the early evolution of our planet: "The birth of Earth was violent and hot. How did such an angry young planet grow and differentiate into the seemingly well-adjusted, mature planet we know today?"[15]

What a spectacle it must have been! A newborn star—our Sun—pulling in vast quantities of gas and dust, heated first by the falling debris and then by nuclear fusion as it reached the size where it could ignite. Meanwhile growing planets swept up additional debris from a disk of gas and dust surrounding the central star—a star that briefly fired great jets of plasma out a few light-years in opposite directions, perpendicular to the plane of the growing planets. Of course, we have no actual pictures of the birth of our own solar system 4.5 billion years ago, but the Hubble Space Telescope has given us wonderful images of regions where stars are forming now, which show us what it must have been like.

Probably the most spectacular region of star birth is the Carina Nebula[16] in the southern skies, a vast stellar nursery where new stars

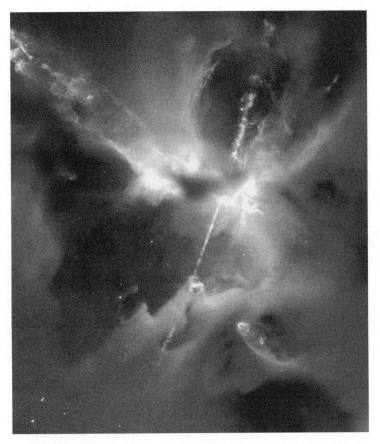

Fig. 2-5. This spectacular image shows the jets of plasma emitted by a newly forming star. Our infant solar system may once have looked like this.

are forming now at the expense of an enormous cloud of gas and dust. Our Sun and Earth may have originated in a mammoth host of newborn solar systems like this, but we do not know because any sister stars would have dispersed around the galaxy in the subsequent 4.5 billion years.

Around the growing and then igniting Sun, there must have been a thin disk made up of bits of solid objects—dust and space rocks—much like the disk that encircles Saturn today, only much larger. Gradually those objects stuck together in bigger and bigger accumulations that eventually became the eight planets we know today as

Fig. 2-6. Part of the Carina Nebula, imaged by the Hubble Space Telescope, where new stars are forming from a great cloud of gas and dust. X-ray studies, which can see through the dust, show over 14,000 stars in this nebula.

well as asteroids and comets. Remarkable computer simulations show us what this process of planet formation may have been like.

As Earth grew larger and larger, there must have been many impacts of nearly planet-size objects—catastrophic impacts that utterly dwarf the Yucatán impact that did in the dinosaurs. If the smaller planet hits the larger one at just the right distance between center and rim, it will tear away a huge portion of the larger planet, sending debris into a disk surrounding the larger one—a disk that will gradually pull together into a satellite. This is currently the favored theory for the origin of our Moon.[17]

The Moon has played a critical part in the human situation, stabilizing Earth's rotation, generating the tides that may have helped marine animals adapt to life on land, keeping most nights from pitch blackness, providing romantic evenings for young couples, helping people construct calendars, and serving as a nearby target for early space exploration and human landings. But a planet with a single large moon is not common—Earth is the only solar system planet that has one—and we might have had no moon, two moons, or a moon that orbited backward, leading to very different human situations, or no humans at all.[18]

The truly giant impact that produced our Moon took place well before the end of the accretion of Earth. You can confirm this with a small telescope, observing that the bright Lunar Highlands are completely saturated with small craters, obviously made by impacts after the Moon came into being. Earth must have received the same treatment after formation of the Moon, but our planet is so active geologically that none of those late-accretion craters have survived here. The post-Moon bombardment was critically important, for the huge impact that created the Moon would probably have dispersed much or all of Earth's water into space. Our present oceans may have been brought in by post-Moon comets, but this is a topic of much current research.

With most of the solar system debris swept up by collision with the growing planets in a hundred million years or so, Earth was on track to develop into a much quieter, better-behaved place. Its tranquility, however, seems to have been interrupted for a while by a resumption of large impacts about 500 million years later. This was the Late Heavy Bombardment we will meet in Chapter 7, and it produced the great craters on the face of the Moon, subsequently filled with dark lava, that we can see with the naked eye. But when that was over, Earth really did settle down into the history of slow continental movements, mountain building, and evolving life that continues to the present time.

We usually take our wonderful Earth for granted because we have known it all of our lives, as have all humans who ever lived. But a Big History sense of its distant past can only leave us amazed and grateful that such a violent and chancy history has given us this perfect place to live.

EARTH

CHAPTER THREE

GIFTS FROM THE EARTH

WE ARE STARDUST . . . *CONCENTRATED BY EARTH*

CARL SAGAN WAS fond of saying, "We are made of stardust." The famous astronomer brought science to life for large numbers of people with his 1980 television series, *Cosmos*, which now looks like an early and very effective presentation of Big History. His point was that with the exception of the three lightest ones—hydrogen, helium, and traces of lithium—all the chemical elements were cooked up inside stars, either slowly as by-products of the nuclear reactions that make the stars shine or suddenly during supernovas, the great stellar explosions that have scattered all the new elements into space. This synthesis of elements in stars was a major point of Chapter 2.

But to a Big Historian like me, coming from a background in geology, Sagan's formulation is incomplete and even misleading. The story does not end with supernovas because chemical elements, dispersed through interstellar space, would be useless for building the world we know. Imagine a spaceship on its way to a distant solar system having an emergency need for silicon to fabricate computer chips. Although the ship is passing through the diffuse gas derived from old supernova explosions and that gas includes atoms of silicon, there would be no way for the astronauts to harvest and use the silicon.[1] The atoms are just too far apart.

43

Yet here on Earth it is easy to find useful concentrations of silicon. We have no trouble finding flint nodules, or sand on a beach, or crystals of quartz in veins, all of which are made of silicon bonded to oxygen—SiO_2. Clearly the role of Earth has been essential in making silicon useful to human beings. The point of this chapter is to explore how Earth does it, in the hope that Big Historians will amend Carl Sagan's idea that we are stardust, and recognize that *we are made of stardust, concentrated by Earth.*

How Earth makes resources useful

How has Earth, made of many different elements all mixed together, separated those elements and turned them into the concentrations of resources we humans find so useful? This is a major concern of geochemists—geologists who use chemistry to study Earth. In understanding this aspect of the human situation, it is useful to recognize that there have been two main phases in Earth's history of sorting chemical elements. The first phase took place at the very beginning, as Earth was accreting. This phase gave us Earth's bulk composition, dominated by the big-four Earth elements—oxygen, magnesium, silicon, and iron—with only small amounts of almost all the other elements but with all the components mixed up together.

The second phase has been going on ever since, with the progressive sorting out and concentrating of elements. Earth has many mechanisms for doing this work, some of them quite complicated, so perhaps it is not surprising that Big Historians have focused on the conceptually simpler nucleosynthesis—the production of new elements in the Big Bang and in stars. The present chapter explores a few of Earth's many sorting mechanisms to give an idea of how Earth works its magic. Our focus will be on the element silicon and on uses of silicon that are important in human history, so we might think of this as a "little Big History" of silicon.

If one had to identify the single most important discovery of geochemistry, it might be this: The *solar system* as a whole is fundamentally composed of a great deal of hydrogen, a moderate amount of helium, and only traces of all other elements as the result of nucleosynthesis during the Big Bang and, later on, inside stars. But *Earth*, on the other hand, is dominantly composed of just four elements—oxygen, magnesium, silicon, and iron (O, Mg, Si, and Fe)—with traces of many other elements, among which the two most important components of the universe and the solar system, hydrogen and helium, are not very abundant. Somehow, Earth has selectively accumulated some of the rare elements of the solar system. Let us see how this great geochemical discovery can be explained and how it affects Big History and the human situation.

Of those four dominant elements, let's focus on silicon because it is the basis of most of the minerals and rocks that make up our planet.[2] Just as carbon is the basis of life, silicon is the basis of rocks.[3] What's more, many rocks carry a record of the conditions under which they were deposited, and geologists have learned to tease out that information.[4] I like to say that rocks "remember" their history.

Another reason to focus on silicon is the critical role it has played in the rise of humanity to enormous technical prowess. The earliest human tools may have been made of wood, of which nothing remains, but the first tools of which we have a good record are made of silicon-based rocks. Moving from just the materials Nature gave us to artificial materials, a critical one has been glass, which basically comes from melting silicon-rich quartz. Finally, our modern, high-tech civilization is dependent on computer chips, which are made in sophisticated ways from silicon.

To see how Earth has concentrated silicon, let us first look at the original accretion of Earth as a planet dominated by O, Mg, Si, and Fe, as we saw in Chapter 2.

EARTH'S FAVORITE ELEMENT (AND OURS, TOO)

Returning to that major geochemical discovery, how can we explain that the solar system is mostly made of hydrogen and helium while

Earth is dominated by the rock-forming elements, O, Mg, Si, and Fe? What processes could account for this wholesale geochemical alteration?[5]

Evidently we're looking for processes that acted very early in the history of the solar system, because today there aren't any large amounts of material entering or leaving Earth, nor is there any sign of this happening in the 4 billion years for which we have a rock record. At the beginning of the solar system, however, 4.5 billion years ago, Earth was growing rapidly, as objects from dust size to planet size accreted and stuck together. There were certainly opportunities during the formation of Earth for selecting some elements to accumulate while excluding others.

Which ones got excluded? Many elements were simply too rare in the solar system ever to have had a chance of becoming important constituents of Earth.[6] Other elements occur primarily as gases—as single atoms or clusters of just a few atoms. These gaseous elements were swept out of Earth's habitat in the inner solar system by powerful outflows of particles from the violent young Sun to accumulate in the giant gas planets farther out—Jupiter, Saturn, Uranus, and Neptune.

Which ones were *not* eliminated? Only *mineral grains*, large enough to see with the naked eye and each containing huge numbers of atoms, were massive enough to resist the pressure of the particles streaming out from the Sun and to stay in the inner solar system where Earth was forming. These mineral grains were composed dominantly of the big-four elements—silicon, oxygen, magnesium, and iron.[7] The most critical one of the four is silicon because its four bonds allow it to link up into enormous networks of atoms, making those mineral grains possible. So we can think of silicon as Earth's favorite element.

Since our focus in this chapter is on silicon, let's go on and see how that element, which was initially mixed up in our planet with oxygen, magnesium, iron, and some minor components, has been sorted and concentrated by Earth into forms humans can use. Our usage of silicon is quite amazing! A number of characteristics distinguish humans from all other animals, and among those features are tools, artificial materials, and computers.

Tools let us do things our natural hands and bodies cannot do, *artificial materials* let us do things impossible with natural materials, and *computers* let us do things our natural brains cannot do. In the rest of this chapter we will look at silicon-based examples from those three categories—stone tools, glass, and computer chips. So silicon is not just one of Earth's favorite elements but one of ours as well. How has Earth concentrated all that silicon?

SILICON AND STONE TOOLS

First let's consider stone tools made of silica. Think of the variety of tools we use today! From simple knives, hammers, saws, and screwdrivers to more complicated tools like violins and pianos, passenger trains, agricultural harvesting machines, and industrial looms, to really sophisticated, high-tech tools like interplanetary spacecraft, laser ranging devices, global-positioning-system receivers, and computer-controlled three-dimensional printers—the list is almost endless and portrays a species with a technological virtuosity that would have been inconceivable even a million years ago.

Primates have been seen to use and sometimes make very simple tools, and some birds may use naturally occurring objects as primitive tools, but the complexity of tool manufacture and use by human beings is quite unique. It is a central part of the human situation. An excellent way to appreciate the vital role of tools in our lives is to read the first few chapters of Daniel Defoe's classic 1719 novel of a marooned sailor, *Robinson Crusoe*, whose survival depends on what he could salvage from his wrecked ship—carpenter's tools, nails, and a grindstone; arms and ammunition; rope, cable, and sailcloth; scissors, knives, and forks—and what he could make himself. Or the modern analogue—an astronaut marooned on Mars, salvaging and improvising whatever tools he could.[8]

How and when did this virtuosity begin? We are unlikely ever to find examples of objects at the transition from found objects to purposely constructed tools because wood, the most likely material, is so perishable. But the record for stone tools is abundant. Archaeologists, who like to divide history into three parts, have long divided the human past into three ages—Stone Age, Bronze Age, and Iron Age—based on the kinds of tools found in excavations.

Until very recently, the earliest recognized stone tools, found in East Africa and dating from about 2.5 million years ago, were those apparently made by *Homo habilis*, whose species name reflects this toolmaking ability. These are simply pebbles with random chips knocked off, giving sharp edges that could be used for cutting. They have been called the Oldowan Industry, after Olduvai Gorge in Tanzania. But now we know of even older stone tools from a place in Kenya called Lomekwi dating from 3.3 million years ago.[9] These Lomekwian tools are contemporaneous with *Australopithecus afarensis*, best known from the fossil called Lucy, and thus predate humans in the sense of genus *Homo*.

Stone-tool industries younger than the Oldowan display increasing sophistication on the part of the toolmakers, with some tools so skillfully and beautifully made that they qualify as works of art as well as practical objects. It seems very probable that human brains and toolmaking abilities increased together in a kind of feedback loop over the last 3 million years.[10]

What made all of this possible is that some rocks, when broken, give very hard, sharp edges allowing humans, with blunt teeth and soft fingernails, to match and exceed the cutting abilities of the animals with the sharpest teeth and fiercest claws.

Nick Toth and Kathy Schick, paleoanthropologists who study stone tools, have founded the Stone Age Institute at Indiana University and are among the pioneers of Big History.[11] They are skilled makers and users of stone tools, and it is a remarkable experience to watch one of them produce an Acheulian hand axe—the main tool used by *Homo ergaster/erectus*, between about 1.5 and 0.5 million years ago. Nick and Kathy point out that stone tools allowed the vegetarian early humans to start eating energy-rich meat when we did not have the teeth or

Fig. 3-1. Nick Toth considers where to strike off the next flake while making a hand axe.

claws to do it naturally. Stone tools gave us the ability to hunt animals larger than we are and to keep away carnivores who were hoping to eat us.

Most rocks do not break with sharp edges or are too soft to hold a cutting edge, so one could never make a useful tool from sandstone, limestone, granite, gypsum, or rock salt. The two best materials for toolmaking are the volcanic glass called obsidian and the sedimentary rock called chert, or flint. Obsidian is rich in SiO_2 but is rare, so let us concentrate on chert, which is almost perfectly pure SiO_2 and fairly common. What is chert, and how does it come to have this very special composition? Or, to pick up the theme of this chapter, how did Earth concentrate the SiO_2 in chert?

Chert commonly occurs as lumps, called nodules, in limestone. Both are sedimentary rocks of biological origin. Limestone is made of the mineral calcite ($CaCO_3$), whereas chert is extremely fine-grained quartz (SiO_2). It is clear from the mineral hardness scale, arranged from softest to hardest (talc, gypsum, *calcite*, fluorite, apatite, orthoclase, *quartz*, topaz, corundum, diamond), that chert will make better tools than limestone.

Most marine organisms make their shells from $CaCO_3$ extracted from seawater, whereas just a few extract SiO_2—sponges and the single-celled floating creatures called radiolarians are the most important. Most freshly deposited limestone mud contains a small amount of dispersed SiO_2, but it can be dissolved by water between the grains of the soft sediment and then concentrated and precipitated in chemically favorable levels, producing beds or nodules of chert, which were greatly prized by our ancestors who made stone tools. This is the first of Earth's silicon-concentrating mechanisms that we will explore.

The famous Stone Age monument of Stonehenge in southern England was built and improved over about 800 years in a place where high-quality chert nodules were particularly abundant in the bedrock limestone. Stonehenge helps us realize that silicon was as critical an element in Stone Age times as it is today, and perhaps we could think of Stonehenge as a very early Silicon Valley.

This history of stone-tool manufacture and use and its probable role in our expanding brains and growing intelligence was made possible because Earth makes chert. Earth has biological and chemical mechanisms for separating silica from the other major elements— iron and magnesium—and from all the other minor elements in our planet, forming beautiful chert nodules of pure SiO_2 from which tools can be made.

SAND FOR MAKING GLASS AND COMPUTERS

Chert is not the only material available for making stone tools with sharp, hard edges—they can also be made from obsidian, a natural glass of volcanic origin. Obsidian forms when molten rock cools so rapidly that there is no time for crystals to grow. So humans have long flaked sharp tools from naturally occurring volcanic glass, but eventually they learned to make artificial glass and use it for other purposes. Glass technology seems to have begun in Mesopotamia or Egypt, about the same time and place people learned to make bronze by mixing copper and tin, as we will see in Chapter 9. So glass and bronze appear to mark the beginning of a history in which humans have learned to make and use an ever more varied array of remarkable artificial materials.

Think for a moment about what glass does for us. Glass windows allow us to have light inside our houses, cars, trains, and airplanes during the day and lightbulbs to illuminate them at night. Glass gives us waterproof pitchers, bottles, and drinking glasses, and it allows us to see in mirrors and through corrective lenses. In telescopes, microscopes, and myriad other scientific devices it has allowed us to understand our world. It insulates high-voltage power lines, provides fiber-optic cables, and gives us monitors and touch screens for computers. In mosaics and stained-glass windows it has been a medium for great art.

Almost all glass made today is based on melting and then rapidly cooling silica, SiO_2, with smaller amounts of other elements added to give it desirable properties. The point here is not the details of how glass is made, but rather the question of how Earth has concentrated the silica we use for glass manufacture. This will lead us to a deeper understanding of Earth's virtuosic ability to separate and concentrate chemical elements, for the silica used in glassmaking was concentrated in a completely different way from the silica used for making stone tools. Fortunately we are not dependent on chert nodules to make glass, for chert is not very abundant. On the contrary, there are enormous supplies of silica in that most common of geological deposits—plain old sand.

For millennia, glassmaking was the main use people had for quartz sand, but now it fills another vital role. In the early twenty-first century we live in a new silicon age, in which this element, the critical ingredient for our Earth itself and once the basis for Stone Age tools, is now again of critical importance as the foundation for the computer chips that pervade almost every aspect of our technology, our lives, and our civilization. The ancient Silicon Valley of Stonehenge has been replaced by the new Silicon Valley in California and similar centers of high technology elsewhere.

The element silicon combines so readily with oxygen to make

quartz (SiO_2) and other silicate minerals like olivine (Mg_2SiO_4) that uncombined native silicon metal is almost never found in Nature. But it is silicon metal that is used in making computer chips. And so chemical engineers have developed industrial processes to remove the oxygen from the SiO_2 in sand and make pure silicon metal. In this case, "pure" means *extremely* pure—as much as 99.999999999% pure!

PLATE TECTONICS AND HOW
EARTH MAKES QUARTZ SAND

Most sand is composed largely, or almost entirely, of little grains of quartz. This is a part of the human situation, familiar to anyone who ever walks on a beach or through the sand dunes. But it is surprising at first, when you think about how Earth concentrates elements, because during the original accretion of Earth there was no way to produce quartz. As we have seen, there were mechanisms to concentrate magnesium, iron, silicon, and oxygen in the newly formed Earth. With abundant Si and O, you might think that lots of quartz, SiO_2, would have formed, but that is not the case. Instead, there was so much Fe and Mg around that all the Si and O were combined with Fe and Mg to make silicate minerals other than quartz—minerals like olivine.

It has taken Earth billions of years to produce the great deposits of quartz sand that we use for making glass and computer chips. The story of how this happened is really remarkable, and at the risk of some oversimplification, I'd like to place it in the framework of the theory of plate tectonics—the great, unifying theory of Earth that emerged in the 1960s and 1970s. This discussion can thus serve also as an introduction to plate tectonics, which will be central to the next two chapters.

"Tectonics" is the study of the large-scale geological features of Earth—continents, ocean basins, and mountain ranges—and the word comes from the same root as "architecture"—in this case, the architecture of our planet. A "plate" is a large segment of Earth's rigid outer layer, going down about 70 miles, with its upper part being continental crust in some places and oceanic crust in others. It might have been better to call the plates "caps," for they are like curved caps fitting on top of the spherical Earth, but the name "plate" has stuck. Most plates contain both continental and oceanic crust, like the North American Plate, which includes the North American continent and about half of the North and Central Atlantic Ocean.

The heart of plate tectonics theory is the realization that each plate is essentially rigid and does not deform very much, but each one moves relative to every adjacent plate, so that most of the deformation of Earth's surface takes place at "plate boundaries." At the first kind of plate boundaries, called "transform faults," one plate slides horizontally past another, as along the San Andreas Fault of California, where the Pacific Plate is moving northwest relative to the North American Plate. Although this kind of movement produces dangerous earthquakes, it plays little or no role in the generation of sand. But the other two kinds of plate boundaries are critical for quartz sand production.

The second kind is called a "spreading ridge," and at this kind of boundary, two plates are moving apart. For example, the North American Plate includes the western half of the North Atlantic Ocean floor, while the Eurasian Plate includes the eastern half. The plate boundary runs along the midline of the North Atlantic, and there the deep-Earth rocks are rising very slowly, and the ocean floor is spreading, with new oceanic crust, made of basalt, continually forming at the midline and adding on to the North American and Eurasian Plates, like two conveyor belts, spreading slowly apart in both directions. This idea was first conceived in 1960 by Harry Hess, my thesis supervisor at Princeton, as we shall see in Chapter 9.[12]

Although most of the spreading ridges of the world's oceans are in the deep sea, there is one place where spreading takes place above sea level—on Iceland, an island of basalt flows, basalt volcanoes, and fractures where spreading is taking place and the island is slowly widening. This is how oceanic crust is formed.

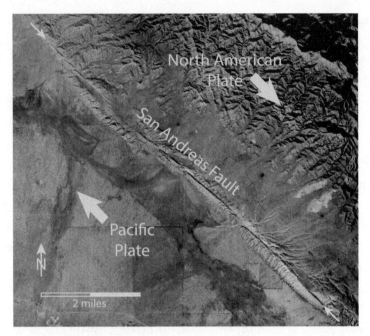

Fig. 3-2. The San Andreas Transform Fault crosses the Carrizo Plain between Los Angeles and San Francisco. The Pacific Plate, here including part of California, has moved more than 300 miles northwest relative to North America and continues to move today.

The third kind of plate boundary is a "consuming margin." Since new oceanic crust is forming at spreading ridges and Earth cannot get bigger, old oceanic crust must be eliminated. So it is "subducted"—it sinks down and returns to the underlying mantle at consuming plate boundaries. Something happens to the subducting oceanic crust—we are still not sure exactly what—but molten rock forms above it and erupts from great chains of volcanoes, like the Andes Mountains all along the western side of South America and the Cascade Range of Oregon and Washington.

Continents ride along on their moving plates, so if the ocean crust between two continents is subducting, eventually the continents will come together in what is called a continental collision. But since continental crust is buoyant, it does not subduct smoothly, like ocean crust, so a continental collision slowly but intensely deforms the leading edges of the two continents, and a mountain range is formed. This is the origin of ancient mountain ranges like the Appalachians and the Alps, which we will visit in Chapter 5. It is also the origin of the greatest of modern mountain ranges—the Himalayas—caught between the colliding continents of India and Asia.

This has been an extremely brief overview of plate tectonics. But what does plate tectonics have to do with the way Earth makes quartz sand? Remember that rocks are solid objects that change very little over geologic time. But at two kinds of plate boundaries—subduction zones and continental collisions—rocks get heated up and they may melt, forming the molten rock called magma. And when rocks melt, all kinds of changes can take place.

Again, with much simplification, one change that happens is that as the magma cools and solidifies, the first minerals to solidify are dense minerals, poor in SiO_2. Being denser than the magma, they sink and the remaining magma becomes richer in SiO_2. When that remaining magma solidifies, it forms minerals richer in SiO_2 than

Fig. 3-3. This narrow trough in Iceland was dropped down between fractures formed where the North American Plate and the Eurasian Plate are spreading apart.

Fig. 3-4. Mount Saint Helens, in the Cascade Mountains of Washington above a subducting slab of oceanic crust, as it erupted in 1980.

Fig. 3-5. These pale-colored Dolomite Mountains in northern Italy are made of sedimentary rocks deposited on top of Italian crust before it was driven up over European crust in the continental colli-sion that built the Alps.

were in the original rock before it melted. Plate tectonics thus works like a giant chemical processing plant—or better, two giant chemi-cal processing plants—gradually producing rocks richer and richer in SiO_2. Deep-Earth rocks contain about 44% SiO_2, ocean crust has about 50%, the volcanoes above subduction zones have about 60%, and the granite that results from continental collisions contains about 75% SiO_2. That last value is high enough for quartz grains to crystal-lize, and indeed, granite usually contains about one-third quartz.

So now we have the quartz crystals, but they are locked up in hard granite, far below the surface. How does Earth get them out and turn them into pure quartz sand?

From the perspective of our lifetimes, measured in decades, or even from the perspective of all of human history, mountains seem like permanent features of the landscape. But from the geological point of view, mountain ranges are ephemeral. They are born and grow slowly to great heights, and then they go into decline and are slowly eroded away so that eventually little or no topography remains. Even the deep roots of mountains, with their masses of granite, rise slowly to the surface and are exposed by erosion. As a result, granite is a common rock in what we can now see of ancient mountain ranges. At this point, the role of plate tectonics in the production of quartz sand is finished.

It is time for weathering to take over and purify the quartz. Chemical weathering at Earth's surface is the same mechanism that produces soil. During weathering by soil acids, all the minerals in granite, except quartz, decay and are converted to clay, especially in hot, wet climates, because they are not chemically stable at Earth's surface. Clay minerals occur as very tiny grains, which are easily carried away by water and wind, leaving the quartz behind. The grains of quartz are extremely stable and endure pretty much forever, accumulating in sand dunes, river channels, and beaches. Over time, Nature cements these sand deposits into the rock called sandstone. The purest sandstones may have an SiO_2 content of nearly 100% and contain almost nothing but quartz.

And so, through different parts of the plate tectonic process, followed by intense weathering, Earth has completed the job of making quartz on a planet that originally had none at all. The volumes of quartz sandstone produced by Earth are quite astonishing. One such sheet of sandstone that covered North Africa and Arabia between 530 and 440 million years ago would have been enough to bury the United States including Alaska in quartz grains to a depth of a mile![13] And now in many places around the world these great sand deposits are quarried for the quartz used to make glass and computer chips.[14]

EARTH'S MATERIALS AND THE HUMAN SITUATION

We have seen Earth's different ways of accumulating the silica we use for making stone tools, glass, and computer chips. But that barely scratches the surface of the rich array of mechanisms by which our planet has produced the concentrations of the resources we use. Almost every chemical element gets enriched by Earth in one or more ways. But concentrations of elements and of resources like fossil fuels are not evenly distributed—different resources are concentrated at different parts of Earth's surface. This has had a huge effect on the human situation and on human history.

At the present time, the irregular distribution of petroleum has enormous consequences for economics, politics, and international relations—and this is what people in the future will look on as *their* history. Regions rich in oil, like parts of the Middle East, are both the beneficiaries and the victims of these natural riches; places that lack oil suffer and benefit in different ways. Quartz sand is lacking in Hawaii, which is made of basalt with no quartz, but abundant in the Sahara. Yet there are different kinds of quartz sand, and Saharan countries rich in sand dunes must import sharp, angular sand from glacial debris in northern countries because the rounded grains from desert sand dunes are ineffective for sandblasting!

From a Big History perspective, we can draw the lesson that not only are we stardust, *concentrated by Earth*, but also that the irregular distribution of those concentrations has played a fundamental role in channeling the path of human history. The occurrence patterns of valuable natural resources like chert for making stone tools, precious metals like gold and silver, and fossil fuels hint at another critical part of the human situation—wealth and its unequal distribution among nations and among people.[15] The geological history that has produced those irregular occurrences certainly does not explain all of the economic aspects of human history, but it has been and continues to be a major factor influencing the economic history of humankind.

CHAPTER FOUR

A PLANET WITH CONTINENTS
AND OCEANS

WATER AND LAND IN THE HUMAN SITUATION

ON CHRISTMAS EVE of 1968 the first humans to leave Earth behind them orbited the Moon. Awestruck by the contrast between the bleak and lifeless Moon beneath their spacecraft and the beautiful Earth in the far distance, they radioed back to home the words from Genesis: "In the beginning God created the heaven and the Earth . . . and God saw that it was good." Through their images, we humans had our first look at a place with no water whatsoever. In contrast to the Moon, we live on a planet that is abundantly supplied with water. This is an absolutely fundamental part of the human situation, for without water there would be no humans and no life, and Earth would be just another dead object orbiting in empty space, like Mercury, Venus, and the Moon.

But Earth is more than a water planet, for it also has dry land—whole continents of land, and islands as well—and this is also a fundamental part of the human situation, for we humans are land animals. An Earth without dry land might have given rise to animals as intelligent as we are, but it is hard to imagine them building anything like our machines and cities and electronic communications in an environment as corrosive as seawater. As dry-land creatures, we live on several continental masses and a lot of islands, and the geographical configuration of continents and oceans has been a major determinant of human history.

The geography of the continents outlines the patterns of the kingdoms, empires, and republics that have been the large-scale actors in human history. The topography and climate of continents have controlled the pattern of settlement and the lines of communication throughout history; resources are distributed in an irregular way across the continents; and land warfare is carried out on a geographical chessboard. The geography of the oceans has determined routes of exploration, trade, and migration and has set the stage for naval warfare.

Reaching back further into human history, we emerged as a species in the isolated, tropical continent of Africa, and our human nature, with all its blessings and curses, is a product of that origin. When our ancestors left Africa about 60,000 years ago, their route was constrained by the geographical pattern of land and water on the planet. There were apparently two pathways out of Africa, one through the Sinai Peninsula, acting as a bridge to the Middle East, and the other across the narrow straits at the southern end of the Red Sea, leading to Arabia, a moat made narrower than today because of the lowering of sea level when much water was locked up in glacial ice on Canada and Scandinavia. Glacial sea-level lowering also made possible the migration of the Asian people who came to populate the Americas. As humans slowly spread around the world, their routes were channeled along coastal plains and river valleys and impeded by mountain barriers.

The overall shapes of the continents may have mattered as well. In *Guns, Germs, and Steel,* Jared Diamond has argued that the east-west elongation of Eurasia and the north-south alignment of the Americas had a major influence on their societies and peoples, which in turn had great consequences when Europeans and Americans came into contact after 1492.[1] Even that contact, with all its violence, disruption, and disease, was played out over decades and centuries in a geographical setting determined by the patterns of continents, oceans, and islands.

This dependency of history on the geography of continents and oceans is not a new concept, of course. Historians and others who look at maps are well aware of it. But human history has been so brief—a few thousand to a few hundred thousand years at most—that the geographical setting has remained essentially fixed. Glaciers have come and gone, with the result that sea level has fallen and risen, coastlines have been submerged, fjords drowned, and bays silted up, but the basic pattern of continents and oceans has not changed significantly in comparison with the dramatic geological changes that take place on a time scale of tens or hundreds of millions of years.

The Big History viewpoint, however, offers a major insight unfamiliar to many people. Over geological history, which is a million times longer than written history, the arrangement of continents has changed radically, and the reconstructed continental maps of hundreds of millions and billions of years ago are quite unrecognizable. Human history has unfolded on a map that is a single frame in a movie in which continents have sailed very slowly about on the globe in kaleidoscopic patterns, sometimes accumulating into giant supercontinents and at other times dispersing into separate continents like we have today.

Before we start looking at those maps of the ancient world, it is worth thinking about time scales for a moment. We will see maps with labels like 200 million years ago and 700 million years ago, and at first those will just seem like incomprehensibly long times past. To make sense of those enormous times, here is how geologists think about them: Written human history goes back about 5,000 years, but Earth history goes back about 5,000 *million* years. So the scale of Earth history is a million times longer than that of human history.

To make sense of geological times, it therefore helps to say that 5 million years in Earth history is like 5 years in written history—it is quite recent. The dinosaur extinction, 66 million years ago, is comparable to 66 years ago in human history—something many people can remember. The start of the fossil record about 500 million years ago is as remote in Earth's past as the Renaissance is in the human past. This is an easily cultivated habit that really helps in thinking about the history of our planet.

PORTUGAL, SPAIN, AND THE MAPPING OF THE CONTINENTS AND OCEANS

In our twenty-first-century technological world we know the coastlines of all the continents and islands with great accuracy and we can examine them on satellite images available online, zooming in to look at every coastal tree and bush. There are no geographical secrets or mysteries in an age of satellite imagery. But until recently this was not the case.

Six centuries ago, people's knowledge of the shapes and positions of continents and islands was extremely limited and unreliable. Early medieval descriptions of distant places were based largely on rumor and fantastic tales.[2] By the late Middle Ages, Islamic scholars like ibn Battuta were traveling the huge Muslim world, describing the places they saw, and Marco Polo recounted a long journey across Asia, although no travelers at that time were compiling accurate maps.

The only quantitative information on coastlines was the list of measured latitudes and estimated longitudes in Claudius Ptolemy's *Geography*,[3] compiled a millennium earlier and full of serious errors, though in retrospect it is amazing what Ptolemy achieved. However, by the fifteenth century, travelers' tales and ancient authority began to be replaced by purposeful exploration and better measurement of positions.

In the early fifteenth century the Ming Dynasty Chinese sent great fleets of large ships around the Indian Ocean, perhaps for exploration as well as to collect tribute, but for reasons that are still not clear the Chinese abandoned exploration about 1425 and turned inward. And so, at just that time, it was left to tiny, impoverished Portugal, on the far edge of the Iberian Peninsula, at the most remote extremity of backward Europe, to begin the systematic exploration that would eventually knit the world together.[4]

Why Portugal? Facing the Atlantic and cut off from the rest of Europe by hostile Castile, which was the dominant power in what was to become Spain, Portugal was not at that time involved in the wars that distracted France, England, and the Spanish kingdoms.[5] Instead, it was exploration that offered the Portuguese an outlet for the crusading spirit that animated the age. And perhaps most important, they were driven by a remarkable and complicated man of genius, Dom Henrique, known in English as Prince Henry the Navigator.[6]

Henry's motivations are not fully known—perhaps to satisfy a lust for gold, to outflank the Muslims of North Africa, to link up with the legendary Christian king called Prester John, or to find souls to convert to Christianity—but he used the financial resources of an order of monastic knights under his control to send his squires out in tiny ships called caravels with orders to push ever farther south along the African coast. It was a slow process—Cape Bojador, previously the most distant point known to Europeans, was passed by the explorer Gil Eannes in 1434; Cape Verde, where the Sahara gives way to rich equatorial vegetation, was reached in 1444; El Mina, in contact with the rich gold fields of West Africa, in 1471, a decade after Prince Henry's death; the Cape of Good Hope in 1488; and on to India in 1498. By that time the Portuguese had made spectacular progress in mapping the continents and oceans.

Historians can document the Portuguese voyages, and scientists can admire the discoveries, but it is difficult in an age of satellite images to appreciate the emotional experience of venturing into the fearsome unknown in tiny ships in a time of unquestioned religious belief and superstitious terrors. I find I can experience those fears vicariously through Samuel Taylor Coleridge's late-eighteenth-century poem, *The Rime of the Ancient Mariner*. Historical accuracy and geographical precision are not the point of the poem, but the beliefs, superstitions, and terrors of early sea voyages come through with clarity:

The very deep did rot: O Christ!
That ever this should be!
Yea, slimy things did crawl with legs
Upon the slimy sea.

About, about, in reel and rout
The death-fires danced at night;
The water, like a witch's oils,
Burnt green, and blue and white.

And some in dreams assured were
Of the spirit that plagued us so;
Nine fathom deep he had followed us
From the land of mist and snow.[7]

Spain joined the enterprise late, when Columbus stumbled on the Americas in 1492. Contrary to popular belief, Columbus did not demonstrate that the world was spherical—that had long been known to scholars—and the story that he defended the spherical Earth before flat-Earth professors at Salamanca is simply wrong.[8] The main intellectual contribution of Columbus was the erroneous belief that the diameter of the globe was much smaller than the correct value, which by then the Portuguese and the Salamanca professors knew quite well. This mistake let Columbus believe that he could reach Asia by sailing west, which his critics knew was ridiculous. The only thing that saved Columbus was the unexpected presence of the continents of the Americas, which demonstrates what scientists know about discoveries—that it is sometimes better to be lucky than to be smart.

Within a hundred years of Columbus, the Spanish and Portuguese had explored much of the continents and oceans previously unknown to Europeans, and the map of the globe was really taking shape. But the human cost was appalling, with lethal plagues like small pox wiping out much of the original population of the Americas and grim diseases like syphilis erupting in Europe. Vast numbers of enslaved Africans were transported to the Americas. European countries conquered empires in the Americas and later in the rest of the world, discovered great accumulations of gold and silver and became rich, while the previously dominant cultures of Islam, India, and China fell behind. We are still living with, and trying to deal with, the consequences of the voyages of discovery. It is clear that the existence of the American continents was enormously important for the human situation, and it was geological history that gave us those continents and made it possible for Asians to populate them long before the Europeans came.

In 1949, the French historian Fernand Braudel, then the leader of the influential *Annales* school of historical writing, published his masterpiece, *The Mediterranean and the Mediterranean World in the Age of Philip II.*[9] Braudel emphatically rejected the concept of history as merely a sequence of human events—what has been called "just one damn thing after another." He insisted that historians take a much longer view, the *longue durée*, and to begin by understanding in detail the Mediterranean geography that has controlled or influenced human history. All of the first and half of the second of his two volumes are therefore dedicated to this setting of the scene for the reign of the great Spanish king.

Braudel appears to have drawn the line at considering the geological history that has produced the landscape of the Mediterranean, but what was known in his time was frankly not very interesting. Sadly, Braudel was writing too soon. Geologists did not yet know about continental drift and plate tectonics and had no understanding at all of the dramatic geological history that has produced the Mediterranean Sea, the Iberian Peninsula, and the whole pattern of continents on the globe.[10] It is only in the last few years that the outlines of the geological story have come into focus, and so in this chapter I have tried to write what Braudel might have written had he done so 50 years later.

The revolutionary discovery
that continents move

Given our lifetimes of a century at most, a time span in which nearly all geological change is imperceptible, we naturally think of the physical world as a fixed and permanent stage on which our lives, and history itself, are played out. But beginning with Nicolaus Steno in seventeenth-century Italy, geologists gradually came to appreciate that the physical world undergoes dramatic changes over very long spans of time—that mountains slowly rise and are gradually eroded away.[11]

Ironically, however, geologists themselves long considered continents and ocean basins to be permanent features of Earth's surface. Mountains might come and go, the geologists thought, but the continents on which they were built have not moved around. They found plenty of markers that demonstrated vertical motions, like shallow-water fossils on mountaintops, but not finding markers for large-scale horizontal movements, they concluded that no such movements had taken place. Looking back today, we see that this was a logical fallacy, because absence of evidence is not evidence of absence.

In fact, however, a marker for large-scale horizontal movements was there all along in the remarkable fit of the coastlines of South America and Africa. If indeed the two continents had once been side by side, they have moved thousands of miles away from each other since then. It was the Portuguese and Spanish explorers who first traced those coastlines, and as early as 1502, on a Portuguese map called the Cantino Planisphere, we can sense the fit of Brazil into the recess of Africa. By 1570 the fit all along the Atlantic was quite clear and Abraham Ortelius, a Flemish cartographer who made a map in that year, actually pointed out the coastline match and argued that "America . . . was not sunk . . . so much as torn away from Europe and Africa, by Earthquakes and flood."[12] It astonishes me that Ortelius got it right less than a century after Old World people started mapping the New World!

In 1912 the German meteorologist Alfred Wegener presented a detailed theory of continental drift, starting from the coastline fit.[13] He reassembled the continents into a supercontinent he called Pangaea, meaning "all the lands," and showed many cases of geological features now sundered that are continuous on a map of Pangaea—like matching the picture as well as fitting the shapes of jigsaw-puzzle pieces. By the late 1920s, however, geologists had evaluated Wegener's continental drift and rejected the evidence and the theory, largely because no one had proposed a mechanism for getting the continents to move.[14] Geologists went back to their comfortable view of fixed continents.

Wegener was finally vindicated, posthumously, three decades later. Marine research during and after the Second World War made it clear that the oceans are geologically very much simpler than the continents, which means that they are geologically young. Princeton geologist Harry Hess started the plate tectonic revolution in 1960, when he suggested that as continents move apart, new ocean floor grows and widens between them, thus explaining why the oceans are younger and geologically simpler than the continents.[15]

Throughout the 1960s and 1970s, with one exciting discovery after another, Hess's idea of an expanding ocean floor developed into the much broader and more powerful theory of plate tectonics. In plate

Fig. 4-1. Harry Hess with a Wayúu Indian in the Guajira Peninsula of Colombia in 1963.

tectonics, continents are seen to ride around on a dozen or so "plates," each moving relative to its neighbors, from which it is separated by plate boundaries. As we saw in Chapter 3, plates grow at boundaries like the Mid-Atlantic Ridge; they subduct—they sink back into the deep-Earth mantle—at consuming boundaries like the Andes, and they slip past each other at transform boundaries like California's San Andreas Fault. This theory is now universally accepted by geologists, who recognize that Earth makes no geological sense whatsoever except in the light of plate tectonics.

I was one of Harry Hess's last graduate students at Princeton, and although I did not contribute to developing the theory of plate tectonics, I had a ringside seat. I got to watch as one piece of the puzzle after another fell into place with wonderful discoveries, demonstrating beyond a doubt that continents move. What I remember most about that time was the almost unbearable intellectual excitement. It was a full-blown scientific revolution, and no geologist who lived through it will ever forget it.

CYCLES IN EARTH HISTORY

One of the pleasures of Big History is that it lets us back off and think about the whole sweep of the past and about the character of history, which is hard to see in the usual kind of detailed studies that geologists, paleontologists, archaeologists, astronomers, and historians like to do. The last chapter of this book will deal with how history unfolds, but the theory of plate tectonics offers an advance peek at that question.

History is sometimes thought of as an interplay between long-term trends, which Steven Jay Gould called *time's arrow*, and repetitive patterns, or *time's cycle*.[16] Time's arrow is unmistakable in many aspects of Earth history, and in plate tectonics we see a clear trend in the gradually increasing complexity of the geology of continents. During the building of a mountain range, rocks are deformed, sometimes almost beyond recognition, are cooked and may even melt, producing volcanic eruptions at the surface and granite bodies at depth. One of the special challenges and pleasures of doing geology is figuring out the sequence of events that has produced the rocks we can study in mountain ranges. Over time, young mountain structures deform older ones, and the cumulative result after 4.5 billion years of this trend is a geological complexity in the continents that continues to challenge and delight geologists.

Being so used to complexity, geologists were amazed to discover, in the 1960s, the very simple geology of ocean basins, and we came to understand that this is the result of time's cycle. When a continent breaks into two pieces that slowly separate, a new ocean is formed between them, starting its history from zero, and before long it will be subducted back into the deep Earth. This complete cycle of ocean creation and destruction is repeated again and again, and no ocean lasts long enough to become very complicated geologically.

Actually we can recognize three different cyclic patterns in plate tectonic history. The most local is the *geologic cycle*, which geologists have understood for two centuries. In the geologic cycle, mountain ranges are built, destroyed by erosion, and new mountains built elsewhere. The geologic cycle is what you read from the rocks at any particular place, and it was obvious long before anyone knew about plate tectonics.[17]

Now, in the light of plate tectonics, we recognize that the geologic cycle is a manifestation of the opening and closing of ocean basins—a fundamental historical pattern that we call the *Wilson cycle*, honoring the Canadian geologist, J. Tuzo Wilson, who first recognized it.[18] The opening of a new ocean basin creates two new continental margins, where thick masses of sediment can be deposited. Later, the ocean basin is subducted and the continental margins collide, deforming the new sediments and pushing up a mountain range that is slowly eroded away, while somewhere else a still younger ocean basin will eventually open up.

However, backing off and looking at the Wilson cycle in the broadest context, geologists now recognize that it is part of a broader *supercontinent cycle*, in which sometimes pieces of continent are dispersed around the globe, as they are today, and sometimes most or all continental crust is gathered together into a supercontinent, the most recent of which we call Pangaea.

Today much active research is focused on understanding the history of the few supercontinent cycles that have taken place in Earth history.[19] There is reasonably convincing evidence for a supercontinent older than Pangaea, one that we call Rodinia, whose assembly and breakup involve fascinating geological stories,[20] and there are even hazier indications of one or maybe two still older supercontinents. Geologists are so fascinated by the supercontinent history they are uncovering that they seldom take the time to explore the implications for human history. So let's do that now, and in the spirit of Big History and of Fernand Braudel, it will be interesting to tie the supercontinent cycle into the Portuguese and Spanish explorations.

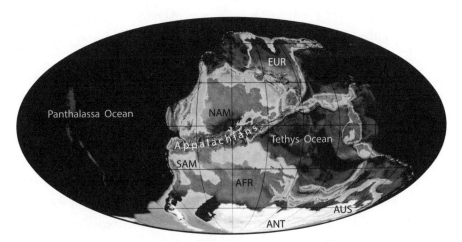

Fig. 4-2. Ron Blakey's reconstruction of the Pangaea supercontinent, 300 million years ago. Pangaea, meaning "all the lands," included both continent above sea level (dark grey) and continent submerged beneath shallow seas (light grey). Deep ocean is black and included the huge Panthalassa Ocean, meaning "all the oceans," and the Tethys Ocean, a gulf that indented Pangaea on the east. The Appalachian Mountains (including the Variscan Mountains in Europe) were formed by a great collision between the northern and southern continents during the final assembly of Pangaea. White is a southern ice cap.

THE FATE OF PANGAEA AND
THE FUTURE OF EXPLORATION

Before its breakup, Pangaea was shaped something like a big pie, with one wedge-shaped piece missing on the eastern side, forming a great embayment that geologists call the Tethys Ocean. A great deal of complicated geological activity took place in the Tethys, which geologists are trying hard to understand.[21]

Pangaea endured for a very long time, from about 320 to 200 million years ago, and then it began to break up into today's dispersed continental pieces, which are such a fundamental part of the human situation. Iberia, the region that would become Spain and Portugal, was located within Pangaea near the western apex of Tethys, and it gradually became isolated as successive breakup lines fragmented Pangaea. By about 115 million years ago the blocky shape of Iberia was recognizable, showing the coastlines that allowed first Portugal and then Spain to become the pioneers of global exploration.

In the Treaty of Tordesillas in 1494, just two years after Columbus found the Americas, the Pope divided up all of the world to be explored and colonized. The Portuguese got everything east of a rather poorly defined line of longitude in the Atlantic Ocean, and the Spanish got everything west of that line. This treaty, which today seems like incredible papal arrogance, makes historical sense in the light of the Christian Reconquest of Iberia, which continued for four centuries and only ended in 1492. During the Reconquest, to avoid conflicts among the Christian kingdoms of Portugal, León, Castile, and Aragon, papal treaties would divide up, ahead of time, the Muslim territories the Christians planned to conquer, and Tordesillas simply expanded that practice to a global scale.[22]

The pattern of fractures that broke up part of Pangaea about 125 million years ago left a large eastward projection of South America, extending east of the Tordesillas line, and that projection was to become Brazil. This combination of geological and human history explains why Portuguese is spoken in Brazil, and why this language of a small, isolated European country is now the sixth most important language of the world in terms of number of speakers.

There was yet another way in which continental movements would influence human history much later on, explaining why Portugal started the explorations almost a century before Spain joined in. Within the western Mediterranean there has been a complicated dance of pieces of continental crust so small that they can only be called microcontinents. The most obvious example is the islands of Sardinia and Corsica, which formerly lay along the southern coast of France before rotating to their present north-south alignment.[23]

Recent studies have shown that the Alborán microcontinent, another of the fragments, moved westward and collided with the southeastern side of Spain, pushing up the great mountain range of the Sierra Nevada.[24] It was in these high and easily defensible mountains that the Emirate of Granada, the last remnant of Muslim Iberia, was able to hold out for 250 years after the rest of the peninsula fell to the Christian Reconquest. The Portuguese, facing only gentle terrain in the southwestern part of Iberia, not yet reached by the slow, westbound collision, completed their part of the Reconquest by AD 1250. The Spaniards, confronted by the nearly impregnable Emirate of Granada in the Sierra Nevada, were not able to finish their part of the Reconquest until the epochal year of 1492. By then the Portuguese had been exploring the oceans for the better part of a century. It was the moving Alborán microcontinent that determined where the high mountains would be and, thus, who would start exploring first. I like to imagine that Fernand Braudel would have loved this insight into the combined geological and human history of the Mediterranean!

THE DESTRUCTION OF LISBON

By 1755 the glory years of fifteenth- and sixteenth-century Portuguese exploration were long over, but Lisbon remained an extraordinary city, still the capital of a global empire and packed with architectural

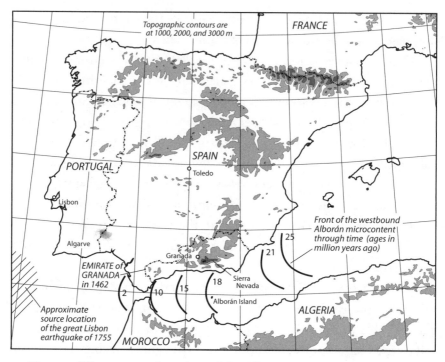

Fig. 4-3. The westward motion of the Alborán microcontinent has affected the history of Spain and Portugal. First, by pushing up the Sierra Nevada, it made the Muslim Emirate of Granada difficult for the Spanish Christians to conquer until 1492, the same year they began to explore the world with Columbus. Portugal more easily conquered the low-lying Muslim south by about 1250 and thus could start exploring almost a century before Spain. Second, the great Lisbon Earthquake of 1755 may mark the front of the crust being pushed up in front of the Alborán microcontinent.[25]

treasures and all the archives and mementos of the great discoveries. And then on November 1, All Souls' Day, disaster struck. Although no instruments existed to measure it, the Lisbon Earthquake is thought to have been the greatest to hit Europe in historical times, exceeding 8.5 in magnitude.

Between earthquake shaking, tsunami flooding in the low areas, and fires above flood levels, the city of Lisbon was effectively leveled and much of Portugal ruined in 1755. The death toll must have been enormous. Wonderful buildings were destroyed. The secret archives of exploration that would have answered so many questions are gone, and scholars will never know what they once contained.

In addition to the physical destruction in Portugal, the Lisbon Earthquake overturned the eighteenth-century intellectual world of Europe. For the religious it was incomprehensible—why would God destroy all the churches in the city and on a very holy day? For philosophers it was equally puzzling—how could Nature be so malignant in a world that seemed to be beautifully designed for human use? Voltaire's *Candide* reflects the disillusionment of European thinkers who had believed they lived in the best of all possible worlds—a belief specifically shaken by the Lisbon Earthquake and the horrific Seven Years' War that immediately followed (1756–1763).

For geologists today the Lisbon Earthquake has a different significance. It appears to have taken place in the ocean floor just southwest of Iberia, and a remarkable 2013 paper by an international team led by Portuguese geologists suggests that the Lisbon Earthquake just might mark a critical turning point in the supercontinent cycle.[26] Perhaps, they propose, the Lisbon Earthquake may mark the new front of the westbound microcontinent that pushed up the Sierra Nevada and made the Emirate of Granada so difficult to conquer. Going even further, they suggest that this may be the very beginning of subduction within the expanding Atlantic Ocean. Perhaps this is the start of conversion of an expanding ocean to a contracting one and, thus, the changeover point in the supercontinent cycle.[27] Maybe the Atlantic will eventually disappear, and a new supercontinent will form in a few hundred million years! It is a fascinating speculation, but sadly no geologist we know will be around to confirm it.

DID SCIENCE BEGIN WITH THE PORTUGUESE DISCOVERIES?

A few years ago I had the good fortune to develop a friendship with Henrique Leitão, a Portuguese historian of science who has a doctorate in physics and is at home in both the humanities and the sciences.

We found ourselves asking the question whether modern science could be said to have begun in the fifteenth century with the Portuguese voyages, taken as a revolution in geology, a hundred years before Copernicus initiated the revolution in astronomy and physics in 1543, which is the usual date given for the beginning of modern science.

Certainly any such claim is a matter of opinion, but it seemed worthwhile to us to make the case, because the Copernican date is so universally accepted that few people would ever think that modern science might have started earlier or in geology instead of in astronomy and physics.

It seemed to us that we could make a pretty strong argument.[28] Although the Portuguese explorers were medieval people at a time when the concept of science did not yet exist, they were doing the kinds of things scientists do today—asking questions about the world they lived in and going out to seek answers. And they were asking questions about Earth, which is what geologists do now.

To go and seek those answers, explorers developed new ways of traveling in the unfamiliar and hostile environment of the vast Atlantic Ocean—the little, maneuverable explorers' ships called *caravels* and the big transport ships called *naos*—just as NASA and planetary geologists do in the hostile environment of space today. They invented long-distance navigation and instruments, like the nautical astrolabe that was developed by the Spanish-Portuguese Jewish astronomer and mathematician, Abraham Zacuto, making that navigation possible. They realized that mathematics was the key to navigation, and perhaps the greatest European mathematician of the sixteenth century was the Portuguese scholar Pedro Nunes, who is Henrique's central interest in the history of science.

The explorers made systematic, quantitative measurement of winds, currents, the deviation of the magnetic compass, and the configuration of coastlines, which they recorded on increasingly accurate maps, just as modern geologists do. And out of those observations came discoveries that today's geologists could recognize as belonging to our field—the great sweeping loops of wind and currents over the oceans, the pattern of magnetic declinations, the seven climate belts of the globe—two polar belts, two temperate belts, two low-latitude desert belts, and the intervening equatorial belt rich in vegetation— and, of course, the matching coastlines of South America and Africa that later led to the theories of continental drift and plate tectonics.

Henrique pointed out that if this was the first scientific revolution, it was unlike later ones in that it was not carried out by intellectuals but instead by ordinary people—by sailors and by captains of little boats. In fact the new view of the world they were uncovering was resisted by the intellectuals of the time, who came from a thousand-year tradition of reliance on authority.

It was the Portuguese sailors who broke the medieval reliance on ancient authorities like Aristotle and Ptolemy. Aristotle had said there were five climate zones, because he did not know about the equatorial belt of vegetation, but the Portuguese explorers found there were seven. Ptolemy's map had shown the Indian Ocean as an enclosed sea unreachable from Europe by ship, but the Portuguese went there by ship and corrected Ptolemy's map. And, of course, the rejection of authority and the reliance on observation and experiment are major characteristics of modern science.

One final observation resonates with me in particular. As I mentioned earlier in this chapter, the plate tectonic revolution of the 1960s and 1970s was characterized by intense excitement and by delight in the discoveries that were being made. That seems to have been the case during the Portuguese voyages as well. On a visit to Lisbon, Henrique showed Milly and me the delightful stone carvings at the Monastery of the Jerónimos, dating from the time of the Portuguese explorations. Ropes carved in stone are everywhere, with exotic birds, flowers, and seashells, an armillary sphere representing Earth and the heavens, and even a ship's cat stretched out on a square knot. These and other Portuguese carvings in the style named after King Manuel are playful, fully secular, and completely different from the somber contemporary religious and heraldic sculptures of Spain.

It is clear from the Manueline carvings that people in Portugal around 1500 were delighting in the discoveries of their explorers. Henrique told us about reports that when an explorer's ship was

Fig. 4-4. Carvings in Manueline style from the Monastery of the Jerónimos in Lisbon, conveying the excitement of the Portuguese voyages of discovery.

sighted entering the harbor of Lisbon, people would rush down to the dock to see and hear what marvels had been found. Today, in a time of almost daily discoveries, it is hard to imagine the contrast this made with the limited, static worldview of the medieval centuries.

Somehow it seems appropriate that the revolutionary discoveries the Portuguese were making as they explored the coastlines of the continents in the fifteenth and sixteenth centuries triggered the same kind of excitement that geologists experienced 500 years later, as the plate tectonic revolution finally explained why those continents have the shapes they do.

CHAPTER FIVE

A TALE OF TWO MOUNTAIN RANGES

MOUNTAINS IN BIG HISTORY

THE PREVIOUS CHAPTER spoke in passing of the mountain ranges that decorate the continents. Although continents are larger in scale than mountain ranges, somehow mountains are more conspicuous and impressive. There is no place on Earth where one can stand and appreciate an entire continent, and although we can now do that on satellite images, the Americas or Australia seen from space look like maps—interesting certainly, but not emotionally affecting. To stand at the foot of a great mountain range and look up at its towering peaks or to travel into the heart of the mountains and be surrounded by wild and beautiful landscapes is an immediate and moving experience. Mountains are a basic part of the human situation, both physical and emotional.

Mountains are an excellent topic for Big Historians, since they are of fundamental interest to historians both of Earth and of humanity. To geologists, they are primary features of the planet, each with a fascinating history that illuminates the supercontinent cycle, and they also offer extensive exposures of rocks that record the history of Earth before the mountain range was formed. The Apennine Mountains of Italy, for example, were formed a few tens of millions of years ago, but they expose rocks dating back to almost 200 million years ago that have let us understand much of what happened in that long interval, including the great extinction 66 million years ago.[1]

For historians of humanity, mountain ranges have been critical barriers to communication and migration. The Himalayas and the Alps have sheltered Indian and Italian civilizations, though not without incursions and invasions that sometimes have marked major turning points in history. Today, when mountain ranges are easily crossed by airplanes and increasingly by great tunnels, we may forget what formidable controls they placed on history until less than a century ago.

Any mountain range would be geologically fascinating, but some have less connection to human history (the Transantarctic Mountains come to mind!) and are thus more difficult to relate to Big History. Other ranges have played central roles in human history. We could have focused on the role of the Andes in South American Incan or colonial history or its Wars of Independence; or on the role of the Zagros Mountains of Iran from the birth of civilization in Mesopotamia to the wars of the late twentieth and early twenty-first centuries, or the Rocky Mountains in the exploration and settlement of the Canadian and U.S. frontiers, or on the Urals in Russian history or the mountains of Central Asia in the long interactions between steppe nomads and settled agriculturalists. Two fine examples are the Alps and the Appalachians, which will be the focus of this chapter.

The Alps are forming even now because of a relatively minor continental collision between Italy and Europe, and they rise as a great barrier between those regions, a barrier that has been of fundamental importance to European history as far back as we have any record of that history. Seven modern nations control parts of the Alps—France, Switzerland, Liechtenstein, Germany, Austria, Slovenia, and Italy—so there are Alpine languages and place names from the Romance, Germanic, and Slavic families; a single mountain may have different names in different languages. And this is just the beginning of the complexity, both human and geological, of this wonderful mountain range.

The Appalachians formed about 300 million years ago in the much greater continental collision between Laurentia and Gondwana that assembled Pangaea—the most recent supercontinent, as shown in Fig. 4-2. But being much older than the Alps, they have been worn down by erosion and are visually less dramatic. The Appalachians entered written history much later than the Alps, but they played a critical role in first impeding and then allowing the thirteen American colonies to break out of their constricted geography along the eastern coast of North America and to expand into a continent-scale country that in some ways dominated twentieth-century history.

In keeping with the Big History approach, let us look at our two mountain ranges first from the viewpoint of historians, then of travelers and artists and, finally, of geologists.

How historians look at mountain ranges

Scholars of written history have a time frame that goes back a few thousand years at most—less than that, in the Alpine part of Europe, and very much less for the Appalachians in North America. In this brief span of time the mountains have changed very little. The great ice sheets that covered Scandinavia and Canada were already long gone at the opening of written history, and the main changes since then have been very gradual erosion and the slow fluctuations in the Alpine mountain glaciers—which at the present time are shrinking fast.

Because of these very slow changes, historians can only see the Alps as a permanent stage set on which history has played out—an unchanging geographical obstacle separating the northern Europeans—Germans, Celts, Scandinavians, and Slavs—from the Italians. In fact, the Alps are barely visible in most historical writing, perhaps mentioned as a barrier but rarely described or analyzed in any detail, because most historians spend their research time in libraries and archives, not in the mountains.[2]

For a historian, the Alps have influenced or controlled great movements in history, separating language groups and religions, with their passes funneling trade routes, military campaigns, and pilgrimages. From Hannibal crossing the Alps to attack Rome, to Caesar crossing them in the opposite direction to conquer Gaul, to Alaric the Visigoth crossing the Alps to sack Rome in AD 410, to Henry IV crossing them in winter to plead forgiveness from Pope Gregory VII in 1077, to Napoleon crossing to first display his military genius in Italy in 1796, to the campaigns of World Wars I and II, the Alps have frequently and strongly channeled the course of European history.[3]

Archaeologists have a longer vision of Europe, encompassing the waxing and waning of the great ice sheets, the resulting rises and falls of sea level, and the associated changes in the climate of the continent over the last several hundred thousand years. Those times saw the spread first of *Homo erectus* across Europe, then the dominance of the Neanderthals, and finally the arrival of our own species, *Homo sapiens*, the only humans still alive. Until recently paleoanthropologists could go little beyond speculation in trying to picture how these different kinds of humans interacted with each other—did they coexist, interbreed, or fight to the death?—but recent advances in molecular genetics are answering some questions and raising new ones.[4]

Curiously, the Alpine glaciers recently gave us an intimate look back much further into European prehistory than anyone could have dared hope for. In 1991 some Alpine hikers came across a body frozen in the ice. An investigation revealed that this was not a modern accident but a truly ancient murder mystery. The man in the snow had lived more than 5,000 years ago, had been crossing the Alps on foot, and was killed by an arrow. It is hard to imagine that we will ever know exactly what happened to this ancient mountaineer, who has come to be known as Ötzi, but his body and his clothing and equipment, deep frozen in the glacier for all those millennia, have provided the rarest and most improbable glimpse of the life of Europeans of so long ago.[5]

The Appalachian mountain range entered written history much more recently, with the Spanish explorers of the sixteenth century and the gradual establishment of English colonies along the Atlantic coast of North America after 1600. By the time the colonies rebelled against England in 1776, they were effectively hemmed in between the coast and the Appalachians, and their leaders, including George Washington, considered this a major risk to the future of the new country.[6] If the lands on the other side of the mountains were to be controlled by England, France, or Spain, the position of the colonies would be precarious indeed. Today the Appalachians seem a minor obstacle compared to the Rockies, but at the time they presented a formidable barrier. In the next chapter we will see how the landscape produced by a complex geological history allowed the young United States to overcome that barrier and eventually to expand all the way to the Pacific.

Ötzi and the excavations of archaeologists can carry us back to a human world very different from the one we know today, but it is only the truly long chronology of geologists that allows us to see mountain ranges as something almost like living entities—born, growing, maturing, and fading away in the mists of nearly endless time. But before we look at the Alps and the Appalachians in the perspective of deep time, let us first consider mountains from two other very different points of view.

HOW EARLY TRAVELERS LOOKED AT MOUNTAIN RANGES

In our world of today, mountain ranges are treasured for their natural beauty and as places for hiking and climbing in the summer and skiing in the winter. Great painters and photographers like J. M. W. Turner in the nineteenth century and Ansel Adams in the twentieth century have captured the beauty of the mountains in iconic images, and every four years the Winter Olympics celebrate the greatest mountain athletes. For those who are not artists or skiers, a trip to the mountains can soothe the mind and heal the soul. This benign view of mountains is so widespread as to be almost unquestionable today.

But only a few centuries ago mountains were seen in a completely different light. My favorite description, dating from 1657, vilified mountains as "Nature's Shames and Ills" and as "Warts, Wens, Blisters, Imposthumes" upon the otherwise fair face of Nature.[7]

Up until maybe the eighteenth or nineteenth century, travelers approached mountain ranges with dread, and it is not difficult to understand why. Imagine being a traveler trying to cross the Alps from Germany to Italy in the Middle Ages, with no decent map and no guide. Imagine that you have taken the wrong path at an unmarked trail junction and have gradually realized that you are lost in a snowy landscape of rocks and crags. Imagine that you have climbed up a long, steep, and fading trail to a jagged pass where you can finally see over a ridge that you thought would lead down to a village, and instead all you can see is another canyon and another ridge, and another ridge after that. And now the afternoon light is starting to fade; as the sun goes down behind yet another ridge the increasing chill presages a frigid night. There is no help, no place to go for shelter, and you may not survive until morning.

This must have been the nightmare of the early traveler trying to cross the Alps, along with the fear of washed-out trails, snow and rock avalanches, lightning storms, and wild animals. Added to these natural hazards were bandits as well as local lords demanding tolls from those crossing their territories. In an extreme case, Muslim pirates based at Fraxinetum on the southern coast of France raided throughout the western Alps, even capturing, in AD 972, the abbot of the great French monastery of Cluny and holding him for ransom. The difference between this mind-set and the viewpoint of the modern traveler, crossing mountain ranges in comfort, could not be more extreme.

In addition to these very real fears of the traveler, there was an intellectual or spiritual reason for unease when traveling in the mountains. In the Christian West, at least, people believed in a very short historical time scale. The only available information on early history came from the first books of the Old Testament—from the genealogies of the patriarchs, telling who begat whom. For centuries scholars interpreted and reinterpreted these genealogies, fine-tuning a time scale in which all agreed that Earth was only a few thousand years old.[8] On such a young planet, the jagged mountains could only be seen as the wreckage resulting from some terrible catastrophe, and this must have been in the minds of at least the well-read among travelers.

How modern travelers and artists look at mountain ranges

The contrast with our modern appreciation of mountains is dramatic, and indeed this shift in viewpoint must be one of the great, if unappreciated, intellectual revolutions of recent times. What changed? Perhaps three modern developments contributed to a love for mountains replacing fear and hatred—improvements in travel, the Romantic Movement in art, and the discoveries of geologists.

After the decay of the great Roman roads, Alpine travel reached a low point in the early Middle Ages. Then gradual improvements in roads across some of the passes slowly reduced the hazards of travel, but the real breakthrough came in the nineteenth century when railroads were constructed across the mountains.

Even today one can only be amazed at the achievements of the Swiss engineers in laying rails up to and over the high passes. The driving wheels of railroad locomotives will slip if the tracks are too steep. There are three ways of overcoming this problem, all of which were used by Alpine railroad builders. One way is with cog railroads, where a drive wheel on the locomotive and a rack rail, lying between the main rails, have matching gear teeth, and this solution was used to take trains up to high and remote places. The second way is to gain altitude with switchbacks, or more dramatically with helical tunnels that climb slowly while spiraling around inside a mountain wall, emerging close to the entry point but at a higher altitude. There are three of these spiral tunnels on the main line from Zürich to Milan. The third way is to tunnel through from a valley on the north to a valley on the south, eliminating the need to climb to high elevations. The use of these three techniques allowed the Swiss engineers to overcome at last the centuries-long difficulties of crossing the Alps.

In the twentieth century, of course, the development of airplanes reduced the mountains to irrelevance for travelers, for whom they are no longer anything more than an interesting sight to look down on. And now the twenty-first century is seeing the apotheosis of rail travel through the mountains. The new Swiss plan is for two great "base tunnels" to connect northern Europe to Italy, greatly speeding up passenger and freight traffic across the Alps. The Lötschberg Base Tunnel in western Switzerland, 21.5 miles long, has been in use since 2007, and the Gotthard Base Tunnel in eastern Switzerland, 35.4 miles long, opened with a ceremony on June 1, 2016. These two tunnels are so long that they begin and end at low elevation, utterly avoiding the need for trains to climb, so those taking the trip will barely even be aware of the existence of the Alps. To me this culmination of railroad engineering is both awe-inspiring and somehow sad, with the mountains that for so long played a dominating role in European history barely mattering at all. And yet this is simply the latest episode in a trend of transportation improvement that has been one of the ongoing themes of human history.

In the nineteenth century, while railroad building was making Alpine travel easy for the first time, artists of what became known as the Romantic Movement were changing the way we look at mountains. In a reaction against the rigidity of classical art, the intellectual rationalism of the Enlightenment, and the squalor of the Industrial Revolution, these painters glorified Nature and saw beauty where others had seen only dangers and Nature's wreckage.[9]

In the United States, the painters of the Hudson River School in the nineteenth century celebrated the romantic beauty of that river, which we will explore in the next chapter, and the majesty of all the American landscape, even as its pastoral beauty was being overprinted by the cities and factories of the industrial age. Albert Bierstadt, one of the Hudson River School artists who did most of his work in the Mountain West, also painted the Alps, and his painting of the Matterhorn, of which several versions can be found on the Web, is a great Romantic masterpiece.

Perhaps supreme among the Romantic landscape artists, the Englishman J. M. W. Turner traveled and painted the Alps when he was still quite young, shortly after the Napoleonic Wars. With his breathtaking use of color, Turner left us unforgettable images of those mountains—images whose vision of Alpine beauty has inspired generations of artists to the present day. And at the same time, geologists were also changing the way people looked at mountains.

READING MOUNTAIN HISTORY WRITTEN IN ROCKS

The fundamental discovery of the eighteenth- and nineteenth-century geologists was that Earth history has not been brief—not just a few thousand years, but enormously long, going back to an origin that has now been dated as about 4,500 *million* years ago. Together with this discovery came the realization that mountains are not the result of great catastrophes but are the result of long-continued, slow processes that have gradually built up and eroded away the mountainous landscapes we so enjoy today. Mountains, we now realize, are not wrecks—they are sculptures.

Almost everything we know about Earth history comes from rocks, because rocks remember, unlike liquids and gases that forget. Most people are interested in plants and animals that move and grow, not in rocks that just lie there, unchanging. But it is their inert quality that makes rocks such good recorders of history.

Different kinds of rocks remember different kinds of history, and we find them all in mountain ranges. Sedimentary rocks like limestones and sandstones remember the settings where they were deposited—coral reef or streambed or desert dune or glacial moraine—and geologists can identify those settings in much detail. Rocks also remember some of what has happened to them since they were deposited—how they have been folded or broken by faults. If they have been heated or deeply buried, they turn into metamorphic rocks and remember the pressures and temperatures and stress regimes deep within the mountain range. If they are heated enough they may melt and possibly carry up to the surface fragments of really deep rocks that we could not otherwise examine, telling us about conditions in the roots of the mountain range. One of my students recently started an essay with the exclamation, "I love rocks!" It's an acquired taste but a richly rewarding one for those who also love history.

The early geologists had no way to obtain dates in years for the rock record of Earth history, but they could work out the sequence of events using stratigraphy—the study of layered sedimentary rocks. The key was to realize that younger rocks rest on top of older ones, unless there has been some complicating factor. This is the "principle of superposition" discovered in Tuscany by Nicolaus Steno in the 1660s.[10] Geologists also learned to use fossils to correlate rocks— to recognize layers that are of the same age—from one outcrop to another, or from one country or continent to another.[11]

After learning to read Earth history and realizing that this history has been enormously long, geologists began a continuing effort to date *in years* the episodes in Earth's past. After precious little success in the nineteenth century but with growing virtuosity in the twentieth century, we now have at our disposal a wide variety of methods to give ages in years—always with some specified uncertainty—for many kinds of events from the most recent back to the very origin of Earth.[12] With this ability to date geological events securely, geologists are in the process of building up an ever more detailed history of our planet. It is a fascinating chronicle, covering many, many aspects of Earth's past. One of the most interesting aspects is what we have learned about the evolution of mountain ranges like the Alps and the Appalachians.

Rather than saying "dates in years," it would be better to say "dates in million-years." Written history is about 5,000 years old, but Earth is roughly 5,000 *million* years old, so geologists think of "million-years" as the basic unit of time. If we encounter a date like 540 million years ago, which is the beginning of the abundant fossil record, we think of it as equivalent in antiquity to something 540 *years* ago in human history, like the Portuguese voyages of discovery. At first this seems strange, but in fact we do it all the time. For example, we change our units of length when using inches to measure a piece of furniture but use miles to measure a trip to another city. Changing our units of time from years to million-years is the secret to appreciating Earth history.

But it is very fortunate that Nicolaus Steno discovered the principle of superposition and thus invented geology in the geologically simple part of Tuscany. Had he tried to do so in the almost infinite geological complexity of the Alps there would have been no chance of success whatsoever, and no geologist today would ever have heard of Steno.

How to make a mountain sandwich

By the time serious geological exploration of the Alps began, with the work of the Swiss geologist Arnold Escher von der Lindt in the mid-nineteenth century almost 200 years after Steno, geology was sufficiently advanced for Escher to realize in 1840 that he was looking at one of those complicating factors—an exception where older rocks lie on top of younger ones. In the Swiss canton of Glarus, running across a row of high peaks was a sharp line—what geologists call a contact, which separates two different bodies of rock.[13] From fossils, Escher demonstrated that the rocks above the Glarus contact are Permian in age; we know today that they are about 280 million years old. Below the contact the rocks are of Eocene age, about 40 million years old. Clearly the principle of superposition is violated in Glarus, but what is going on?

After much debate, the question was finally answered,[14] and the answer put geologists on the track toward understanding mountain ranges, an understanding that is still being deepened and enriched today. The Permian rocks above the Glarus contact were not deposited in that position; instead they were deposited to the south, and much later they were driven northward—driven up and over the Eocene rocks below the contact. The Glarus contact was thus recognized to be a *thrust fault*—a shear surface where older rocks have been pushed up and over younger rocks. This sort of thing happens when rocks are driven together in compression, and the Alps have become a kind of type example of a compressional mountain belt. In the 170 years since Escher's discovery, geologists have recognized that thrust faults are ubiquitous features in compressional mountain ranges.[15]

A greater discovery about thrust faults was to come. It turned out that the Glarus thrust was just a minor thrust, and much bigger ones were lurking in the Alps. In the later nineteenth and early twentieth centuries geologists were learning to recognize the settings in which rocks were deposited, and they were able to distinguish shallow-water continental-margin rocks from those of deep-water oceanic settings. It gradually became clear that the Alps have three main components, stacked up vertically, with a northern European continental margin at the bottom, an Italian continental margin at the top, and in between them a layer of rocks that originally formed the crust and sedimentary filling of an ocean—an ocean that has completely disappeared! The contacts separating these three components of the Alps are enormous thrust faults, and the compressional nature of thrust faults showed that the vanished ocean—the Tethys Ocean of the previous chapter—once lay between northern Europe and Italy and was squeezed out of existence as Europe and Italy were driven together.

This is a beautiful example of two different kinds of scientific evidence leading to the same conclusion. Fitting the continents together to reconstruct Pangaea leaves a wedge-shaped piece of ocean that geologists call Tethys (see the map of Fig. 4-2). Tethys is gone now, but its oceanic rocks are found high in the Alps. Clearly the history of Earth has been more complicated and interesting that anyone could have ever imagined!

Fig. 5-1. Looking east toward Austria from near Klosters in eastern Switzerland. The snowy peaks are Italian continental-crust rocks, thrust northward over rocks of the Tethyan seafloor, which in turn are thrust northward over rocks of the European continental crust, out of sight, deep below this landscape.

To appreciate the discovery of the vanished Tethys Ocean, I like to think of the Alps as a sandwich. The lower slice of bread would be the ancient continental margin of Europe, the upper slice the continental margin of Italy, and the filling of the sandwich would be the crust and sedimentary filling of the Tethys Ocean. This gives a nice mental picture, but then you realize that Nature makes a sandwich very differently from the way you do. You probably make a sandwich by laying out a slice of bread, spreading the filling out over it, and placing another slice of bread on top. If you wanted to make a sandwich the way Nature does it, you would lay two slices of bread on the table some distance apart, smear the filling (peanut butter and jelly, perhaps) all over the tabletop between the two slices of bread, and then you would slowly push the slices of bread together, mounding up the peanut butter and jelly, and finally push one slice over the other, gooshing the peanut butter and jelly between them. It is not the ideal way to make a sandwich, but it is Nature's way, clearly recorded in the Alps!

What about the Appalachians? In terms of elevation they are not nearly as impressive as the Alps. The highest peak in the Appalachians is Mt. Mitchell in North Carolina, 6,684 feet high, less than half the height of Mont Blanc, the highest Alpine peak at 15,781 feet, and the Alpine peaks are far more rugged and still have active glaciers. This is simply a matter of age: The Alps are still deforming, rising, and being attacked by active erosion to give the rugged topography, but the Appalachians stopped deforming 250 million years ago and their former high peaks have been seriously lowered by erosion.

When we look at their extent, however, we find that the Appalachians are far more impressive. The Alps are only 600 miles long, while the Appalachians are 2,500 miles long, and with a 1,200-mile continuation in the Variscan Mountains of central Europe, their original length was 3,700 miles, or more than six times as long as the Alps.

Internally, however, the Appalachians have the same general structure as the Alps, with the thrust faults and the sandwich character. In the Alps we understand that the sandwich came from the collision of Italy with northern Europe, but what continent could have collided with North America to make the Appalachians? Today there is nothing but the Atlantic Ocean on the other side.

There was no hope of answering this question until the advent of plate tectonics, the Wilson cycle, and the supercontinent cycle. But now we understand. The Appalachians were produced by the collision of Africa with North America—the collision that built the supercontinent of Pangaea (as again you can see in Fig. 4-2). After that collision, however, Pangaea broke up. Africa moved back away from North America and this opened up the Atlantic Ocean, which is younger than the Appalachians.

So the narrow coastal strip in which the thirteen American colonies were trapped was built by two great events in Earth history—

the Gondwana-Laurentia collision that assembled Pangaea, the last supercontinent, about 320 million years ago, and built the Appalachian Mountains, followed by the opening of the Central Atlantic that broke up Pangaea about 180 million years ago and started outlining the continents of our human situation.

HOW TO MAKE A MOUNTAIN SCULPTURE

Have another look at Albert Bierstadt's Romantic painting of the Matterhorn on the Web. The great pyramid of stone rises a thousand meters, or 3,000 feet, above the surrounding ridges, hidden in the clouds, which are already at a high altitude. For a traveler or a mountaineer the Matterhorn is a colossal mass of rock, yet somehow graceful and delicate, soaring into the sky.

A geologist can appreciate the massiveness and grace but also can understand that the Matterhorn is really something completely different—it is just a tiny remnant of what has been eroded away! The evidence of the erosion is all around with deep glacial valleys, whose glaciers, sadly, are rapidly melting away. The very shape of the Matterhorn, a four-sided pyramid, reflects the headwalls of four radiating glaciers that have cut away massive amounts of rock to leave the central peak.

It is not difficult to look at the Bierstadt painting and visualize all the erosion that has taken place down from the level of the peak of the Matterhorn, like Michelangelo starting with a block of marble and chiseling away at the edges to create a statue. The real surprise is that the erosion of the Alps was done as if by a crazy sculptor who buys an enormous block of marble and chips away almost the entire thing until he has a tiny little miniature sculpture. Alpine erosion did not just start from the level of the peak of the Matterhorn—it started from a level far, far higher than that!

However, it is not that the Alps were ever enormously higher than today; it is just that uplift and erosion have gone on together. To carry the sculpture analogy further than it can really tolerate, it is as if

Michelangelo were working on a magic block of marble that keeps expanding, so he has to keep chipping away constantly to keep it the size and shape he wants.

The Alps are the result of a continual competition between compression driven by the Italy-Europe convergence, pushing the mountains upward, and gravity and erosion tearing them down. If we remember that this competition has been going on for tens of millions of years, it is clear that truly massive amounts of rock have been driven up and carried away in the history of the Alps. Geologists would say that the Alps are at a steady state—the Matterhorn is a temporary feature that will be gone in a few million years but will be replaced by other younger peaks carved from rocks that now lie far beneath today's surface.

Indeed, mathematical models designed to mimic this behavior show that the topography of the Alps is at an evolving and fluctuating balance between uplift and erosion.[16] Much of the debris from that erosion now lies in sedimentary basins flanking the Alps on the north and south. This helps geologists who are now working out the history of uplift and erosion of the Alps in detail, finding times when uplift dominated and the Alps grew higher and times when erosion was dominant and the Alps were worn down.

Eventually, however, convergence between Italy and Europe will come to an end, as the supercontinent cycle shifts to a new phase. When that happens, uplift will also come to an end but erosion will continue, the rugged Alpine topography will gradually be reduced, and eventually the Alps will be worn down nearly flat. But as the mountains become lower, the rate of erosion decreases, and it will take several hundred million years to completely erode them away.[17]

IN THE PREVIOUS chapter we saw that the pattern of widely separated continents, which is such a fundamental part of the human situation, is simply one ephemeral configuration in the long history of the supercontinent cycle. Now we have seen that mountain ranges, another prominent aspect of our situation, are also temporary features of Earth's surface, pushed up where continents collide and worn away when the supercontinent cycle shifts to a new phase. Had humans evolved a hundred million years earlier or later, the continents and mountains of the human situation would have been completely different.

People go to the mountains for pleasure, to travel from one place to another, and to mine the resources we need. But far more human activity takes place in lowlands watered by great rivers, so in the next chapter we will be looking at the geological and human history of rivers.

CHAPTER SIX

REMEMBERING
ANCIENT RIVERS

LANDSCAPES FROM THE TRAIN

THE SUPERCONTINENT CYCLE is driven by the heat of Earth's interior—by the escape of heat coming from radioactive decay and the slow solidification of the iron core. Thus it is driven by Earth's internal processes. The rise of mountain ranges is due to internal processes, but their dismantling by erosion is accomplished by external processes, like rivers and glaciers, driven by the heat of the Sun. In this chapter we will look more carefully at the Earth history produced by external processes, especially by rivers, glaciers, and wind. These are the kinds of geological changes that are familiar to most people and produce the landscapes in which we live our lives.

Landscapes, of course, are everywhere and the choice of examples is endless. I've chosen just one example to explore in detail—a transcontinental swath across the United States. Let's do this by way of a train ride that you might like to take some day. Our journey starts in New York, passes through Chicago, Denver, and Salt Lake City, and ends at San Francisco. The train is an ideal way to see the landscapes across an entire continent. Air travel gives a broad perspective, but you have little idea what you're looking at down below; it all goes by so quickly, and the view may be obscured by clouds. In a car you have to pay attention to the road. But on a train you are right in the middle of the landscape, traveling at a leisurely pace, and you can look out the window as much as you want.

Our Amtrak journey will start on the Lakeshore Limited, riding north up the Hudson River from New York City to Albany, then west along the route of the old Erie Canal to Buffalo on Lake Erie. During the first night, we roll through the lowlands of Pennsylvania, Ohio, and Indiana and into Chicago on the shore of Lake Michigan the next morning. There we change trains and board the California Zephyr, traveling westward through the farm country of Illinois and Iowa, interrupted by the Mississippi River, then in darkness across the Missouri River into Omaha and on through Nebraska. We wake up the next morning approaching Denver, with the Rocky Mountains getting closer and closer.

After Denver, breakfast in the dining car affords us the first spectacular views of the Rocky Mountains, which we wind through all day, following the Colorado River for 200 miles through otherwise inaccessible canyons. In the late afternoon we see part of the Colo-

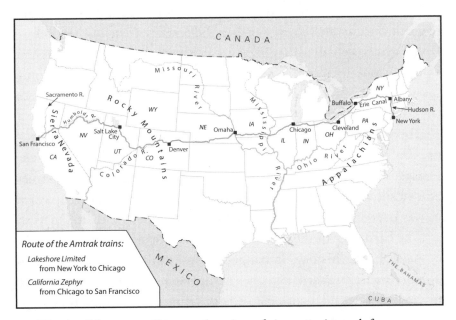

Fig. 6-1. The route of our exploration of rivers on Amtrak from New York to San Francisco.

rado Plateau and cross the Wasatch Mountains, reaching Salt Lake City about midnight. After a nighttime run through the deserts of Nevada, we reach Reno at breakfast time. Then we climb the steep eastern slope of the Sierra Nevada and descend the gentle western slope through Gold Rush country and down into Sacramento. A final leg through the inland delta of the Sacramento River brings us to San Francisco Bay where our journey ends at Emeryville, across the Bay from San Francisco.

There is no more spectacular way to see the landscapes of the North American continent. Looking at the scenery with historical mindedness and a geologist's appreciation of landscape evolution, we will learn about some really remarkable episodes in Earth history and how they influenced human history. Are you ready? *All aboard!*

THE HUDSON RIVER

Pulling out of Penn Station in New York in mid-afternoon, the Lakeshore Limited winds through a maze of underground passages beneath the city and emerges along the eastern shore of the Hudson River, which we will follow all the way north to Albany. For the first 25 miles, looking to the left across the river to the western shore, we see a great cliff of black rock, 400 feet high, called the Palisades, while the contrasting eastern shore is gentle.

Topography like this tells us about the geologic structure, for erosion preferentially attacks weak rock, leaving the hard rock like the Palisades standing high. The cliff is the eastern edge of the Palisades Sill—a sill being a great slab of rock that, as molten magma, pushed its way between sedimentary layers and now slopes gently to the west.

The hard rock of the Palisades cliffs was formed by a really major event in Earth history. It happened 200 million years ago, when North America was still joined to Africa, South America, and Europe in the Pangaea supercontinent. A huge rising plume of hot rock from the base of Earth's mantle, able to flow slowly under deep-Earth pressures and temperatures, reached the base of the crust where it partly

Fig. 6-2. The Palisades cliff, on the western side of the Hudson
River, in a 1903 photograph.

melted. The molten magma was then injected along sedimentary layers, as in the case of the Palisades Sill, or in long vertical fissures,
one of which crosses Portugal and Spain, or erupted onto the surface as basalt flows that can be seen today in Brazil and West Africa.
Geologists call a feature like this one a Large Igneous Province, and
they call this particular one CAMP—the Central Atlantic Magmatic
Province. Before erosion, CAMP was probably one of the largest of
the Large Igneous Provinces, and it coincides in time with the next
to the last of the great mass extinctions.[1] The Palisades Sill is perhaps
the most spectacular manifestation of the CAMP event. Millions of
people live within a few miles of the sill, and many see it or cross over
it every day.

It is common for prominent topographic features to play major
roles in human history, which is the case with the Palisades. In 1776,
near the beginning of the American Revolutionary War, the British

plan was to take control of the long, navigable Hudson River with their powerful navy, thus cutting the American colonies in half. To stop this plan, General George Washington had two forts built on either side of the Hudson—Fort Washington on upper Manhattan Island and Fort Lee in New Jersey—pretty much where the two ends of the George Washington Bridge are today, where their cannon could fire on British ships sailing up the river.

The American plan did not succeed. The British captured New York City, which at that time was confined to the downtown end of Manhattan, and in November 1776, they took Fort Washington. British troops then scaled the difficult face of the Palisades to assault Fort Lee, but Washington, realizing his situation was untenable, had already retreated into New Jersey. It was a low point in the war, and Thomas Paine wrote, "These are the times that try men's souls." Only when Washington crossed another river, the Delaware, on the day after Christmas and won victories at Trenton and Princeton did the American situation improve. As in so many other cases, the topography created by Earth history is inextricably linked to human history.

But back to the Hudson River. As the Lakeshore Limited rolls northward along the eastern shore of the Hudson, we may marvel at the magnificent view. Although not a long river, much of the Hudson is far wider than the Mississippi. And unlike the Mississippi in its mostly flatland setting, the Hudson cuts through magnificent mountain scenery that we can see behind the forbidding, castle-like buildings of the United States Military Academy at West Point. Along the East Coast of the United States, the Hudson is unique—no other river is so wide, so deep, and navigable so far upstream. Why is this so?

The secret of the Hudson is that it is not just a river valley. It is a river valley that was scoured out—deepened and widened—by a tongue of the great Canadian ice sheet during the last great ice age and then flooded when sea level rose from the melting of the great continental ice sheets. This happened about 12,000 years ago. In fact, the Hudson is the only substantial river in the eastern United States that was far enough north to undergo this glacial excavation. As a result, it is navigable all the way to Albany. And not just that, it is *tidal* all that way, so in the days before steamboats, the incoming tide would help carry ships northward against the flow of the river, and the outgoing tide would speed their passage downstream. Altogether an ideal river, the Hudson! It is no wonder the British intended to make it the key to their strategy for defeating the revolt of the American colonies. And that is just the beginning of the Hudson's significance in American history. As the Lakeshore Limited pulls into the station at Albany around dinnertime, we are about to explore what the Hudson contributed to the development of the young United States.

THE ERIE CANAL

Since we are heading west from New York to Chicago, you might ask why we have started by going north for 130 miles, out of our way, to Albany. The answer has to do with the Appalachian Mountains that, as we saw in the previous chapter, formed a serious barrier between the eastern coastal plain of the young United States and everything to the west. The wonderfully navigable Hudson River provides the first half of the only natural route through the Appalachians between Maine and Mississippi. The second half begins when our train turns west at Albany and begins to follow the route of the Mohawk River and the Erie Canal.

Here is a superb example of the connection between Earth history and human history. The Appalachian mountain range is the belt of deformation along the suture where Africa collided with North America to form the supercontinent of Pangaea, about 300 million years ago. Like the crumpled front of a car that has been in a wreck, parallel lines of folded sedimentary rock sweep northeast from Alabama to Pennsylvania—folds that have been detached from the basement rocks along a thrust fault, like a crumpled rug pushed across the floor. In the western part of New York State, the Appalachians change, the folds die out, and to the north, in the Adirondacks and

New England, the deeper rocks have been pushed up and are exposed at the surface.

As a result, there is an east-west belt of lower topography running across the western part of the state, along the line of station stops the Lakeshore Limited will make—Schenectady, Utica, Syracuse, Rochester, and Buffalo. This low belt mostly follows the distribution of weak sedimentary rocks along the northern edge of the Alleghany Plateau, so it is the complicated geological history of New York State that has prepared this natural pathway through the Appalachian Mountains.

The rocks that control the topography are old, but the erosion and deposition that have produced the landscape are very young and are the result of several advances and retreats of ice age glaciers. Most of New York State was covered by ice during much of that time, and the evidence is everywhere. Just south of the rail line are the finger lakes—Cayuga and Seneca and others, about a dozen in all. They occupy deep canyons that were scooped out by fingers of ice and dammed by the debris the glaciers left behind.

For many miles on either side of Rochester, the Lakeshore Limited passes innumerable little rounded hills, elongated north to south, called drumlins. These hills were formed under the ice sheet from debris shaped by the flowing glacier. As we pass through Palmyra, between Rochester and Syracuse, we are in the heart of the drumlin field, and just 4 miles south of the track is a drumlin called Cumorah or Mormon Hill. This is where Joseph Smith said he received the Book of Mormon, from 1823 to 1827, on a set of golden plates supplied by an angel named Moroni. Thus it is here, in a sense, that the Mormons began the journey that led them all the way to Salt Lake City, which we will reach on the train in a couple of days.

Just at that time, work was finishing on the Erie Canal, built between 1817 and 1825, which was of crucial significance in the rise of the United States as a great nation.[2] With independence from Britain established in 1783, George Washington recognized the urgency of finding an easy route across the barrier of the Appalachian Mountains. He feared that the mountains would so separate the frontier from the original 13 states that the frontier region might form its own country or be taken over by France, England, or Spain, trapping the United States in a narrow coastal belt. Washington's attempt to establish a canal over the mountains along the Potomac River was unsuccessful—Earth history had not made that into a practical route.

But the Hudson and Mohawk rivers and the belt of low topography in western New York State made an ideal route, although with obstacles to be conquered, including 500 feet of elevation difference from Albany to Buffalo and many barriers, like Cohoes Falls between Albany and Schenectady and the great Niagara escarpment. Getting the agreement to build the Erie Canal was difficult, and digging the ditch and building the locks in an age of hand labor was herculean, but once the canal was finished in 1825, it changed everything. The agricultural products of the west and the manufactures of the east floated easily along the placid waters of the canal, linking the coastal states and the new interior lands into a dynamic, growing nation. Prices fell, western populations grew, and the Erie Canal–Hudson

Fig. 6-3. A lock on the Erie Canal at Lockport, New York, in the nineteenth century.

River route initiated the growth of New York City from a small town into a great metropolis.

In western New York State, the general line of the canal was followed, a few decades later, by the railroad we are now riding on. The canal itself was widened and in part rerouted in the early twentieth century to make the New York State Barge Canal, which we can see at several places out the train windows, and Interstate 90 was built along the same route. The old, original Erie Canal, very much narrower, is visible occasionally—now just a discontinuous ribbon of still water, flanked by towpaths and shaded by ancient trees—silently remembering those few decades in which it was the great transportation artery that built a new nation.

Ice margin rivers and human history

It is fully dark by the time we reach Buffalo, at the western end of the Erie Canal.[3] Through our first night we roll westward along the shore of Lake Erie, through Cleveland and Toledo, and across northern Indiana, reaching Chicago in the morning. Perhaps we might walk a few blocks and look at Lake Michigan, another of the five Great Lakes that are still a reminder of the glacial age that ended 10,000 years ago. In the early afternoon we board the California Zephyr and head west across the farmland of Illinois.

At dinnertime, the windows of the dining car afford a fine view of the Mississippi River as we cross it at Burlington, Iowa, and during the night we pass over the Missouri River as we come into Omaha, Nebraska. Thus we will have crossed two of the three great rivers of the interior of the United States, making their characteristic Ψ-shaped pattern, missing only the Ohio River because of our northward detour to follow the Erie Canal route. The two arms of the psi—the Missouri in the west and the Ohio in the east—tell a really important story about the landscape evolution of North America. North of these two rivers, almost all of the land has been glaciated; south of it only high peaks were covered with glaciers. The rivers roughly mark the southern limit of continental glaciation. The reason goes back to glacial times. As the glaciers advanced southward, they forced the major rivers—the Missouri and the Ohio—to move farther and farther south. When the glaciers melted back about 10,000 years ago, those rivers were left flowing along the line of the southernmost ice margin, as they still do today.

But the river geography could have been very different, and then human history would also have been different. Steven Dutch is a very thoughtful and creative geologist at the University of Wisconsin-Green Bay. In 2006 he wrote an abstract for a talk at the Geological Society of America meeting, entitled *What if? The ice ages had been a little less icy?*[4] This is an example of counterfactual history, like the book cited in Chapter 2, *What if the Earth had two moons?*[5] There are also books about counterfactual human history,[6] which show how easily the human situation might have been very different.

Steve focused on the last glacial time and explored a scenario in which "the North American ice sheets never extended far below the Canadian border, and the Scottish and Scandinavian ice sheets never merged." He suggested that the Missouri and Ohio rivers would not have been pushed southward to their present courses, and the streams in their present drainage basins would have flowed into completely different river systems. As a result, "the Thirteen Colonies might well have been hemmed in to the west and confined permanently to the Atlantic Coast. There would be no Great Lakes and no Erie Canal. Without the Ohio and Missouri Rivers furnishing easy east-west water transport, American history would have been profoundly different." He goes on to suggest that smaller ice sheets in Scotland and Scandinavia could have led to the British Isles being a peninsula of Europe, with no English Channel, which would have had profound consequences for European history. I consider this a major contribution to Big History, and with Steve's permission his abstract is given in full in the notes.[7] I really cannot think of a better illustration of two of the main themes of this book—how geologic history has influenced human history, and how easily things could have turned out very differently.

DISCOVERING AN ANCIENT, LOST RIVER

After pulling out of Denver in the morning, a dining-car breakfast lets us enjoy the spectacular climb up the Front Range of the Colorado Rockies through a chain of small tunnels, until we cross under the Continental Divide through the Moffat Tunnel, 6.2 miles long and 9,200 feet above sea level. Emerging from the tunnel we immediately pick up the Fraser River, a tributary of the Colorado, which we join in a few miles at Granby. The Zephyr will follow this river most of the day, for 200 miles, through wonderful mountain scenery, including roadless Byers and Gore canyons. In these canyons the river has cut down into really ancient rocks, dating from about 1,700 million years ago and called the Yavapai Belt—rocks that go back far beyond the supercontinents of Pangaea and Rodinia.[8]

In the late afternoon we reach the Colorado Plateau and come to

Fig. 6-4. The Colorado River emerges from Gore Canyon in the Colorado Rockies along the route of the California Zephyr. The rocks in the distance are about 1,700 million years old.

Ruby Canyon and the last we will see of the Colorado River. The walls of the canyon are a beautiful pink color, giving the canyon its name, and are made of a sandstone, called the Entrada, deposited between about 210 and 160 million years ago. It is one of three great Colorado Plateau sandstones that carry a wonderful story of river history.

Geologists have long known that these sandstones were deposited in ancient deserts because their rounded, frosted quartz grains are the result of innumerable collisions between windblown sand grains. The great, downward-sweeping bedding planes, easily visible from the train, mark the fronts of ancient dunes. During the deposition of the Entrada, this part of the western United States must have looked like today's Sahara, and the orientation of the dune-front bedding tells us that the sand was coming from Wyoming and blowing toward the southwest. Recovering the direction of ancient winds was pretty remarkable, but a few years ago the story became more fascinating still, due to a study by Bill Dickinson and George Gehrels of the University of Arizona.[9]

Almost all the sand grains in these ancient dune deposits are quartz, but a very small fraction of the grains are the mineral zircon. Like quartz, zircon is very resistant to the corrosion by soil acids that turn rock into soil, so it lasts indefinitely. But unlike quartz, zircon can be dated because it contains a little uranium, which decays radioactively to lead at a known rate. This in turn gives the age of the granitic source rocks from which the zircon grains, as well as the quartz sand, were derived. Dickinson and Gehrels dated more than 1,600 zircon grains from sandstones of the Colorado Plateau, and to their surprise the ages did not correspond to any possible source rocks in Wyoming or to any known rocks in western North America. The most abundant zircon ages fell between about 1,000 and 1,200 million years old.

The only source rocks of the right age are in the eastern United States—in the Appalachians! They formed by continental collision during the assembly of the Rodinia supercontinent. This ancient collisional zone is the Grenville Belt, which runs all along the eastern margin of North America but is not present in the western part. In fact, the Adirondack Mountains, just north of the eastern third of the Erie Canal, are made of rocks of just that age. Somehow massive amounts of quartz grains, together with the datable zircon grains, were making a journey something like the westbound trip we are taking on Amtrak.

How could quartz from the Appalachians get into sand dunes on the Colorado Plateau? It could not have happened in any recent time because sand carried westward from the Appalachians by either winds or rivers would end up in the Mississippi and would be carried to the Gulf of Mexico. Prior to that, in the Cretaceous, the interior of North America was submerged by the wide, shallow Mancos Sea, and sand grains reaching that sea would stay there.

In the Jurassic, however, the geography of North America was completely different, as Dickinson and Gehrels pointed out. The present situation, with the low Appalachian Mountains in the east, the Mississippi River in the middle, and the high Rocky Mountains in the west, did not yet exist. Instead, Jurassic North America had a long, slow, gentle slope from high topography in the east, uplifted because Africa was breaking away from North America, gradually descending to sea level where the Rocky Mountains are now. A great west-flowing river, whose exact route we can only guess, must have carried vast numbers of Appalachian sand grains all the way to the shoreline in Wyoming, and from there they were blown by the wind to their final resting place in Utah. Who could have imagined, while rolling past the Entrada cliffs in Ruby Canyon, that this is part of the Appalachian Mountains, disassembled and carried 2,000 miles across the continent by a long-vanished river?

THROUGH THE DESERT AT NIGHT

Leaving Ruby Canyon and the Colorado River, the train continues westward through a drab, rolling landscape of dark grey mudrock, the Mancos Shale, harmless in dry weather but a treacherous morass of gluelike mud after a rain. These are the deposits of the shallow Mancos Sea of the Late Cretaceous, about 85 to 70 million years old, which finally brought to an end the transcontinental Jurassic river discovered by Dickinson and Gehrels.

Out the right side of the train, looking to the north, we can see the Book Cliffs, made of resistant, tan sandstone that covers the Mancos Shale. The train follows these cliffs for about 150 miles, raising the question of what geological history they might represent. In a painstaking study led by John Van Wagoner, research geologists of ExxonMobil Upstream Research Company came to understand in great detail that these sands are the deposits of vanished rivers that rose in Cretaceous mountains to the west and gradually filled in the Mancos Sea.[10] Van Wagoner and his friends unraveled a fascinating story of rivers, deltas, and shallow marine sand, sometimes advancing eastward into the sea when pulses of uplift raised the western mountains, sometimes retreating as the Mancos Sea temporarily gained the upper hand. This story is so rich in understanding of Earth history that geologists from all over the world come here to study it, always hoping not to hit one of those rare desert storms that turn the Mancos Shale into a completely impassible barrier.

About dinnertime we reach Green River, a truck-stop town on Interstate 70 where field geologists and river rafters gather in the evening at Ray's Tavern. Here the train turns northward, still following the Book Cliffs. In darkness we climb up into the Wasatch Mountains, cross over Soldier Summit, and descend into Provo and Salt Lake City—the final, unintended destination of the Mormon journey that began along the Erie Canal. We are now in the Great Basin of Utah and Nevada, a large, enclosed desert region from which the rivers, such as they are, cannot escape.

Invisible from the train at night, but prominent in daytime, is a set of horizontal terraces contouring along the base of the Wasatch Mountains behind Salt Lake City. These are ancient shorelines cut by waves on Lake Bonneville, an enormous body of freshwater that occupied the northwestern third of the State of Utah during much wetter times, tens of thousands of years ago. There is an amazing

Fig. 6-5. The horizontal lines along the base of the mountain in this old illustration are shoreline terraces cut by waves on ancient Lake Bonneville, the very much larger ancestor of today's Great Salt Lake, near Wellsville, Utah, 55 miles north of Salt Lake City.

piece of Earth history concerning Lake Bonneville.[11] The lake did not simply evaporate when the wet climate turned dry. Instead it overflowed the drainage divide north of Logan, Utah, about 17,400 years ago and spilled over into Idaho, catastrophically eroding the spectacular outlet channel followed by U.S. Highway 91 through Swan Lake, Idaho. More than 1,000 cubic miles of water poured north through this channel, into the Snake and Columbia rivers and down to the Pacific Ocean. Great Salt Lake is what is left—now evaporated down to a salty residue.

Only one substantial stream still feeds Great Salt Lake, which is the river the train has followed from Provo to Salt Lake City. The early Mormons, steeped in biblical lore, noticed a remarkable analogy: In the Holy Land the freshwater Sea of Galilee feeds the Jordan River, which drains southward into the hypersaline Dead Sea. In Utah, freshwater Utah Lake feeds a river that drains northward into the hypersaline Great Salt Lake. Charmed by the geographically reversed analogy, they named the connector the Jordan River.

If you are awake in the night, you might feel the train speeding along the dead flat, arrow-straight tracks that cross the salt flats west of Great Salt Lake—the evaporated bottom of the western part of former Lake Bonneville. Later in the night, having crossed into Nevada, the train follows the winding path of the Humboldt, perhaps the strangest river in the country. From a source at a modest spring in northeastern Nevada, the Humboldt flows westward, passing through gaps in the many north-south mountain ranges of the Great Basin—looking on a map like a herd of worms crawling north out of Mexico, someone once said. These little mountain ranges are tilted blocks of crust, bounded by faults, for the Great Basin has been pulled apart in extension—California is further away from Utah than it used to be.

The Humboldt cannot escape the bowl-shaped trap of the Great Basin, and eventually it simply evaporates at a dry lake bed called the Carson Sink. But it provides the only easy route from east to west through Nevada. Were the gaps through the mountain ranges cut by an earlier, more powerful Humboldt in wetter times? It seems likely but hard to prove because rivers have the unfortunate habit of eroding the evidence geologists would like to have seen.

The Humboldt provided the route followed in the nineteenth century by pioneers in their covered wagons, then by the transcontinental railroad, and now by Interstate 80. Like the Erie Canal through western New York State, the Humboldt made it possible to open whole new territories of the United States, in this case the State of California.

THE GOLDEN RIVERS OF CALIFORNIA

On our final morning, after a stop in Reno, the train winds up the valley of the Truckee River, climbing the steep eastern slope of the Sierra Nevada. We look down on Donner Lake where in 1846 a wagon train led by George Donner was trapped by an early snowfall. The ugly story of the Donner party lives on in California lore and emphasizes the role of rivers, or lack of rivers, in human history, for there is no through-going river to make an easy route across the Sierra Nevada. Those last 50 miles across the mountains were the most difficult part of the long wagon journey to California.

After Donner Pass and Emigrant Gap, coming down the western slope, the view out the left side of the train shows a smooth surface sloping gently westward, incised by deep canyons. This makes it clear that the Sierra Nevada is a giant crustal block, tilted slightly toward the west. The steep eastern slope we climbed after Reno marks the great fault system that allowed the block of crust to tilt.

The Sierra Nevada, the deserts to the east and south, and the long sea voyages around Cape Horn or across the Pacific protected California from the growing global web of communication and trade for a long time, but those barriers fell to the lure of gold in the mid-nineteenth century. The west-flowing rivers that drain the gentle slope of the Sierra have names that invoke the Gold Rush—Yuba, Mokelumne, Calaveras, Stanislaus. It was here, in 1848, on the South Fork of the American River, that James Marshall discovered gold in the riverbed at a sawmill he was building for John Sutter, whose agricultural outpost at Sutter's Fort, now Sacramento, would be vandalized and ruined by the hordes of gold-seeking 49ers who arrived the next year.

The gold-bearing rivers have a fascinating story in their own right. The first miners to arrive were able simply to pick up nuggets from the streambed and later to wash the sand and gravel down through sluice boxes—wooden troughs with ridges to catch the heavy gold flakes. But the hordes of 49ers quickly exhausted the gold supply in the river gravels and soon discovered where the streambed gold was coming from.

The original source of the gold is in a system of quartz veins called the Mother Lode, which has been mined in deep underground workings. But there were also really rich gold deposits in river gravels, although not the gravels of today's rivers. They were in the gravels of an ancient river system that flowed across the Sierra Foothills region in the Eocene, about 50 million years ago, long before the tilting

Fig. 6-6. Hydraulic gold mining in the Sierra Nevada of California in the nineteenth century.

of the modern Sierra Nevada. Those ancient gravels tell us about a lost river system called the Ancestral Yuba River.[12] The Eocene gold-bearing gravels are buried under younger volcanic rocks, but they are exposed on the sides of the modern canyons. There the miners learned how to extract the gold by washing the gravels down with huge hoses and running them through sluice boxes.

It was a very effective way of mining but an environmental disaster, changing parts of the Sierra Foothills into wastelands like Malakoff Diggins, which can still be seen as a scar on satellite images, a century and a half later.[13] The vast amount of debris washed down by hydraulic mining overwhelmed the Sacramento River and the Central Valley, destroying agriculture, and eventually reached San Francisco Bay, choking parts of the bay with fine sediment. The devastation was so great that in 1884 California passed one of the first environmental laws in the United States, banning further hydraulic mining.

Passing Berkeley, our last view from the train is of San Francisco and the Golden Gate Bridge, looking across the Bay and showing us one last bit of river history. San Francisco Bay is very young. Before the melting of the Canadian ice sheet about 10,000 years ago, sea level was about 300 feet lower, and there was no San Francisco Bay. Instead the Sacramento River flowed in a valley where the Bay is now, through a deep gorge at the site of the Golden Gate Bridge, out across a coastal plain, and into the Pacific Ocean near the range of hills that has now become the Farallon Islands.

And so we reach Emeryville, the end of the line, after a four-day journey of almost 3,400 miles. On our transcontinental train trip we have had a chance to sample some of the wonderful histories of how rivers come to be and how they have affected human history. The great cities at either end of our route owe their prominence to rivers but for different reasons. New York became great because it was the terminus of the Hudson River, the magnificent, glacially scoured fjord that led to the beginning of the Erie Canal, the only natural route across the Appalachian Mountains to the beckoning lands of the North American interior. The speed with which the frontier then swept west across the continent is amazing. San Francisco became great because its wonderful harbor, a drowned river valley, led to the Sacramento River, whose tributaries, rich in gold, attracted hordes of miners, only a quarter century after the completion of the Erie Canal.

Surely it is fair to say that you cannot really understand human history without understanding the crucial role played by rivers. But rivers are simply a given, an unexplained feature of the landscape, unless you also learn about the remarkable geologic history that lies behind each one.

LIFE

CHAPTER SEVEN

YOUR PERSONAL RECORD
OF LIFE HISTORY

"ENDLESS FORMS MOST BEAUTIFUL
AND MOST WONDERFUL"

WITH THESE WORDS Charles Darwin ended his great explanation of the history of life, *The Origin of Species*. But how, as
Big Historians, can we achieve an understanding of something that
is "endless"?

To deal with this problem, let us try an approach I have not
encountered before: Let's take our own bodies as a record of the history of life. As Big Historians, it helps greatly that our main focus is
on the events leading to human beings and human history, for that
allows us to concentrate on just one evolutionary lineage—our own.

Fossils in rocks were the first record of the history of life that scientists learned to read, during the first great age of geological discoveries
in the nineteenth century. In the late twentieth century, fossils were
supplemented by a new source of information about life history—the
genetic record of evolution carried in the DNA of the cells in every
organism.

Fossils and DNA give complementary records of life history, each
supplying information the other cannot. Fossils tell us what an organism looked like, while DNA tells how two organisms are related.
Genetic information from DNA is more easily analyzed and quanti-

fied than the shapes of fossils, but DNA does not survive unaltered in extinct organisms, except for the most recent, like our Neanderthal cousins.[1]

For people in good health, the body works so well that it seems like all its parts simply belong together, like a well-constructed machine. But the record of life history in fossils and DNA tells us that different parts of the body first appeared at very different times, changing through time to be sure, but with older or younger pedigrees. The human body is a palimpsest, put together over billions of years.

Think about what you can easily see in a mirror—a symmetrical body; a mouth with teeth, a tongue, and a moveable jaw; forward-facing eyes; arms and legs; hands with opposable thumbs; hair; a dry, downward-pointing nose; skin that covers your skull, which in turn hides your large and active brain. Which of those parts of your body has the oldest pedigree? To find out, let's explore some of the major features of your body in the order in which they first appeared.

Here is a time line, as a guide to the history we are about to explore:

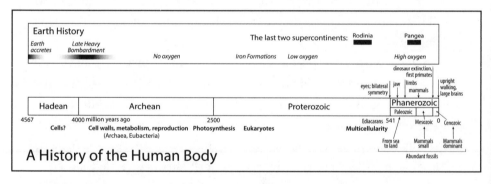

Fig. 7-1. A very simplified history of life, as it has led to our human bodies, showing a few major features of Earth history to provide context.

THE ORIGINS (HADEAN AND ARCHEAN)

The parts of our bodies with the most ancient pedigree are invisible to us. Cells are too small to see with the naked eye and were not discovered until the invention of the microscope. Our bodies contain tens of trillions of cells, differentiated into many specialized kinds—nerve cells, muscle cells, blood cells, skin cells, and so many others.

We do not yet know of any reliable way to determine when cells, and when life itself, first appeared. Surely it was after the last time Earth was largely molten, after the giant impact that produced the Moon, as mentioned in Chapter 2.

Our current understanding is that maybe a half billion years after accretion was essentially over and Earth had quieted down, there was a repeat episode of very large impacts that we call the Late Heavy Bombardment.[2] Impacts from this time are responsible for the huge, dark-colored, basalt-filled craters that are visible to the naked eye on the otherwise light-colored, crater-saturated Moon. Presumably Earth was also blasted during the Late Heavy Bombardment, although no such craters have survived almost 4 billion years of geological history. It is very likely that cells were present by the early Archean, shortly after the end of the Late Heavy Bombardment, and they may even have originated in the Hadean and survived the Late Heavy Bombardment. So every cell in your body must trace its earliest ancestry back to times either just before or just after that episode of intense, renewed impact cratering.

At a minimum, a living cell must be enclosed by a cell wall that partially isolates its contents from the surrounding environment, it must metabolically process materials and energy, and it must be able to reproduce itself. In what environment might cells with these features first have come to be? In a now famous letter to Joseph Hooker in 1871, Darwin imagined life originating in "some warm little pond." Today that quiet origin seems unlikely.

To me the most interesting hypothesis is that life may have first appeared at the mid-ocean spreading ridges we met in Chapter 3, in submarine hydrothermal vents. At these vents, water heated by contact with newly formed oceanic crust at a mid-ocean ridge spews up into the ocean, carrying a brew of elements dissolved out of the scorching hot rocks that only shortly before were part of Earth's mantle. Coming into contact with cold seawater, some of those elements precipitate out as minerals that build up chimneys around the emerging hot water.

A few decades ago, when the origin of oceanic crust by seafloor spreading could only be studied indirectly, geologists were eager to go down and see what was happening there. Fortunately the spreading axis of the Mid-Ocean Ridge is shallower than the deep seafloor and could be reached by the research submersible *Alvin*, operated by Woods Hole Oceanographic Institution. In 1977 Jack Corliss led a group of scientists who did the first dives on the ridge on a segment near the Galápagos Islands. They were about to make a truly great scientific discovery![3]

Probably they were not surprised when they found the submarine hydrothermal vent chimneys, for that is what they expected in this very hot, wet, geologically active place. The first vents found by the Corliss team were not all that impressive, with warm water issuing from little mounds on the seafloor, but later explorers found amazing chimneys rich in metals extracted from the ocean crust by circulating hot water. Chimneys up to 200 feet high have been found, with clouds of black, metal-rich water at temperatures much higher than the sea-level boiling point of water, pouring out in roiling clouds. They called these chimneys "black smokers." Many have been found since, and there are remarkable videos of black smokers on the Web. Clearly these black smokers, or hydrothermal vents, are a major feature of Earth, playing an important role in oceanic chemistry.

But the really astonishing discovery of the Corliss team was that these deep-sea hot-water vents are full of life! They are flourishing oases in the otherwise barren desert of most of the deep seafloor. The team found huge clams, mussels, and limpets. There were also creatures they called dandelions, delicate and other-worldly-looking relatives of jellyfish, unknown to science and unlike anything ever seen before. Most astonishing of all were animals the team called tube worms, sometimes many feet long, living inside flexible white tubes the animals construct.

This was an ecosystem so deep in the ocean that it had never known sunlight. In contrast, almost all known ecosystems at Earth's surface ultimately derive their energy from photosynthesis, but with no light whatsoever, these deep-sea communities get their energy from sulfur extracted by hot water from the rocks of the ocean crust. As oceanographers have continued exploring the Mid-Ocean Ridge, they have discovered many new vents, some with huge chimney systems. They've found other exotic animals, including crabs and fish, adapted to life at the extreme pressure and temperature conditions of the vents. They've also come upon collapsed chimneys where the hot-water flow has ceased and the communities have died.

These deep-sea vent communities were so astonishing and so unexpected that they have forced paleontologists to reconsider fundamental questions about the origin of life itself. Did life begin at Earth's surface, as had always been assumed? Or could it have originated deep in the ocean, at some primordial hydrothermal vent, as a by-product of plate tectonics?

One very interesting suggestion is that life might have originated not at full-blown, extremely active black smokers with their very high temperatures, but rather at older vents with more modest hydrothermal activity.[4] The Lost City Hydrothermal Field in the Atlantic Ocean is attracting much interest as a modern analogue for this possible environmental setting for the origin of life.[5] Charles Marshall, director of the University of California Museum of Paleontology, pointed out in a recent lecture that a vent like Lost City, in early Earth history, would have provided abundant materials, long-lived gentle energy flows, and little spherical holes in rock where proto-cells could have been protected, have shaped their first cell walls, and have gradually evolved the ability to metabolize and reproduce.[6] In addition, the deep oceans would have offered protection to early life from the impacts that may have repeatedly sterilized the surface waters.[7]

Again we have no sure way of knowing whether places like Lost City might have hosted the very first life. And we do not know for sure when the essential features of life appeared, but probably in the Hadean or early Archean. The picture is hazy, of course, but in a very general way it allows each of us to trace our personal genealogy back 4 billion years!

THE LONG INCUBATION
(ARCHEAN AND MOST OF THE PROTEROZOIC)

With the establishment of enclosed and protected cells that can use energy and materials and can reproduce, Earth entered a very long interval of time in which those single-celled organisms diversified and refined their ways of making a living. This long phase of incubation lasted more than 3,000 million years, including much or all of the Archean and most of the Proterozoic. The variety of tiny organisms that emerged is quite astonishing.

It is easy to tell a giraffe from a turtle just by looking at them, but this is not the case for single-celled organisms. Under a microscope they are mostly little balls or tiny elongated sausage-shaped objects, perhaps with one or more whiplike flagella that allow them to move. Once all known as bacteria, we now know from studies of their DNA that there are two great groups: (1) Eubacteria, or true bacteria, and (2) Archaea, the ancient ones. They are as different genetically from each other as either one is from us. Within each group there are many ways of obtaining energy and nourishment—what my Princeton professor Al Fischer called "evolutionary inventions."[8] There are microorganisms that survive by using energy derived from iron, nitrogen, and sulfur, as well as by using solar energy to turn water and CO_2 into living organic matter—the latter, of course, is photosynthesis.

This photosynthesis profoundly disrupted Earth's ecology because its by-product is oxygen, O_2. Although we think of oxygen as necessary for life, it was a lethal poison for the early microorganisms, and we are descended from ones who were able to evolve a tolerance to it. The oxygen revolution is extremely important for the human situation, not only because we breathe oxygen, but also because it was responsible for the great deposits of iron ore on which much of our industrialized civilization depends.[9]

Iron occurs on Earth mainly in two chemical forms—"reduced" (ferrous iron, Fe^{2+}) and "oxidized" (ferric iron, Fe^{3+}). Since the young

Earth did not have oxygen in its atmosphere, most iron at the surface was of the reduced type, Fe^{2+}, and there was an enormous amount of it—remember that iron is one of the big-four elements of our planet. After the appearance of photosynthesizing microbes, their waste oxygen entered the atmosphere where it mostly went into oxidizing Fe^{2+} to Fe^{3+}. In effect, Earth's surface slowly rusted during the Proterozoic, for the mineral hematite, Fe_2O_3, is rust, made of oxidized iron, Fe^{3+}. A critical detail is that ferrous iron is soluble in seawater, but ferric iron is not. So as ferrous iron was gradually oxidized to ferric iron, it precipitated out of the seawater, and accumulated as sedimentary layers rich in iron, forming huge deposits of hematite called banded iron formations. Much of the iron used in industry comes from banded iron formations in China, Australia, Brazil, Africa, Russia, India, and Minnesota.

At some point in that long Archean and Proterozoic incubation, a most remarkable event took place, giving us the kind of cells our bodies are made of. Our cells are Eukaryotes, a third great division of life, along with Eubacteria and Archaea, and they arose when two ancestral kinds of cells found that they could profitably live together, one inside the other, in an arrangement called endosymbiosis. The original host is now the main part of our cells, including a nucleus containing our main DNA, while the partner, or endosymbiont, gives us our mitochondria, which do the energy processing that drives our cellular activity.[10] It is not clear when in the more than 3,000-million-year incubation this event took place, but the two components of our cells have never fully merged, and they keep their separate DNA.

It might seem that we are entirely made of eukaryotic cells, which have completely diverged from Archaea and Eubacteria. But a recent realization is that vast numbers of both Archaea and Eubacteria live in our digestive systems, and this microbiota, or microbiome, is vital to our ability to use food for nourishment. The lack of oxygen in our stomachs and intestines makes them a comfortable environment for these microorganisms, many of whom never evolved to tolerate the oxygen by-product of photosynthesis. That buildup of oxygen in our atmosphere has been called the greatest-ever event of air pollution.

And so the cells of our bodies, including our microbiome, provide a historical record of some of the great events in the history of life during that long, long stretch of time in the Archean and Proterozoic.

CELLS LIVING TOGETHER (LATE PROTEROZOIC)

One of the truly great events in the history of life is dramatically remembered in our bodies, for we are not single-celled organisms. Instead we are intricate assemblies of tens of trillions of cells, specialized into many different kinds and cooperating in complex ways that allow us to function. The origin of multicellularity is not well documented in the fossil record, but probably the first step was for identical, nonspecialized cells to live together as colonies, as the single-celled choanoflagellates can do today. Only later is it likely that specialized cells appeared.[11]

Remarkable little fossil embryos have been found in China and dated to 750 million years ago. These tiny spheres have preserved the various stages of division into multiple cells, and there can be little doubt that these are from multicellular organisms.[12] Then, in sedimentary rocks from late in the Proterozoic Eon (around 600 million years old), paleontologists have found imprints of soft-bodied animals large enough and complicated enough that they were almost certainly multicellular; these are the Ediacarans.

That it took probably more than 3,000 million years for single-celled organisms to be able to live together as many-celled animals suggests that this was neither an easy nor an inevitable step for life to take. In fact it may be an extremely improbable step, and it has been argued that although single-celled life may be common in the universe, multicellular life may be very rare. Paleontologist Peter Ward and astronomer Don Brownlee have called this idea the "Rare Earth" hypothesis.[13]

Eubacteria and Archaea are commonly shaped like simple little balls and sausages, but with the arrival of multicellular animals, different body plans evolved. Some of the oldest have radial symmetry, like sponges, corals, and jellyfish. Our ancestors diverged from this simple geometry and ended up with a bilaterally symmetric body plan. So when you look at yourself, or someone else, and notice the left-right symmetry of the face and body, you are seeing the historical record of a body plan that originated in one branch of life close to 600 million years ago and has endured ever since.

In single-celled organisms, each individual takes care of all its own necessities, from obtaining nourishment, to mobility, to reproduction and many others. When our ancestors learned to live as assemblies of cells, those different functions were divided among specialized types of cells, and that division of labor has continued on to us today.

So there are cells that work in the nourishment trade, from the mouth at one end, through the digestive tract, to waste elimination at the other end. Almost all bilaterians have this complete nourishment system, as we do. It is surprising to think about how much food enters our mouth and passes through our digestive tract—close to 100 tons in a long lifetime! At your next meal, you might think about that number and realize that in processing and utilizing all that food you are using a system whose origin dates back more than half a billion years.

The other main systems that allow multicellular organisms to function must also go back more than a half billion years—the cardiovascular, sensory, nervous, and reproductive systems. There have been great changes in each of these systems through that half billion years, but it is fair to say that none of your multicellular ancestors ever lived without some version of each one.

Because of the rarity of well-preserved fossils of soft-bodied animals, there is little evidence in the rock record for what happened during the early evolution of multicellular animals through the end of the Proterozoic, and we depend heavily on information from DNA. But all this was to change dramatically at one of the key dates in life history, 541 million years ago, the beginning of the Phanerozoic—the Eon of Visible Life.

FROM SEA TO LAND (PALEOZOIC)

Life is abundantly visible in the fossil record beginning about 540 million years ago because of the development of hard parts—in the case of snails and clams that means shells; for us it is bones and teeth. Geologists use this appearance of fossils to mark the beginning of the Cambrian Period. Multicellular animals do not have an absolute need for hard parts, as they have for a digestive tract, and indeed the earliest ones lacked them. But at that milestone moment, hard parts appeared across a wide range of animals. Probably this was part of a natural arms race, and animals that produced hard parts were more likely to survive and reproduce. A common view is that trilobites evolved eyes, allowing them to become much more effective predators, and in response, only prey species with hard parts capable of keeping predators at bay survived.[14]

The effect in the rock record is dramatic: Shells suddenly appear in the sedimentary rocks. Hard parts preserve much better in rocks than soft tissues, and so this looked to the early geologists like a sudden origin of life. Now we understand that there was much life before 540 million years ago, but the evidence is much harder to find.[15] So in your personal document of life history in your body, bones go back to a beginning somewhat more recent than your bilateral symmetry, circulatory system, and digestive tract.

One feature that makes possible the enormous consumption of food by the digestive system is our movable jaw, with teeth for cutting and grinding food. Thinking of the jaw as a historical document, we can trace its origin back to the Ordovician, when our lineage diverged from jawless fishes like the modern lampreys. To feed, the lamprey fastens its mouth onto a fish by suction and uses its tongue to rasp away the flesh of the victim. In the history of life, it is common for a feature that originally served one purpose to find other uses down the road. This is the case with our jaw, a structure evolved for eating, about 460 million years ago, which became central to the development of spoken language about 1 million years ago. Eating with your jaw is a much more ancient activity than using it to tell stories!

Fossil bones give us a window into the early history of other parts of our bodies, like the skull that encloses our brain and the vertebrae housing the spinal chord that brings sensory information to the brain and carries back its commands for movements. And bone fossils also tell us about the history of our arms and legs.

All the history of our bodies we have discussed so far is a record of life in the sea. But as land animals, we are naturally interested in when our ancestors first left the water for the dry land. Single-celled organisms probably got there first, but there is little if any record of that migration. Land plants emerged next, in the Silurian, about 435 million years ago, and thrived once they evolved their own hard parts, which we call wood, enabling them to stand up without the support of the buoyancy of water.

With plants covering the dry land, there was nourishment waiting there for animals, and their migration to the land took place in the next period—the Devonian, about 420 to 360 million years ago. Paleontologists have recently been very successful in finding transitional fossils between fish with bony fins and land animals with legs, and those transitional animals have wonderful names like *Panderichthys*, *Tiktaalik*, *Acanthostega*, and *Eryops*.[16] These are the actors in the early history of the arms and legs that are so critical to our bodies and how we move about and use tools. Fossil bones make it possible to study the fin-to-limb transition in detail, but other parts of our bodies that had to adapt to life out of the water—like lungs replacing gills and reproductive systems that work in dry-land conditions—are not so well documented. Nevertheless it is clear that much of our bodies have been shaped by that long adaptation to the originally hostile conditions on land.

SURVIVING IN THE SHADOWS (MESOZOIC)

The animals who made the transition from sea to land had four limbs; therefore, all their descendants including amphibians, reptiles, birds, and mammals like us are called tetrapods. It is interesting to think of what might have been if an alternative history had taken place. Suppose those transitional animals had had six limbs rather than four. In that case, land animals might have used four limbs for walking, which is much easier than two, and have had two others for manipulating tools like the legendary centaurs. In that case tool use and intelligence might have come much earlier than in the history that actually did happen.

Among the tetrapods, amphibians were most important early on, in the Carboniferous and Permian, because they laid their eggs in water as had their marine ancestors. Reptiles were the next major players because the shells of their eggs made it possible to have their young on dry land. Among the reptiles, dinosaurs appeared in the Triassic and were the dominant large animals during the Jurassic and until the end of the Cretaceous. In a sense, dinosaurs are not entirely extinct, for their descendants, the birds, live on today. And finally came the mammals, giving birth to live young and nourishing them with milk from the mammary glands that give us our name and are part of the historical record in our bodies.

At one time scientists thought of the sequence amphibians–reptiles–mammals as a progression of improvements and that mammals were in some sense the highest type. But evidently someone forgot to tell the reptiles that they were obsolete, for dinosaurs dominated the large-land-animal niches for more than 130 million years. Mammals were around for most of that time but were restricted to small body sizes. What's more, the dinosaurs evolved into the most amazingly varied kinds of animals—moderate-sized, large, and enormous; herbivores and carnivores; walkers, runners, swimmers, and fliers, and with shapes that no one could have imagined until the fossils were found. No wonder dinosaurs are the delight of small children!

In the meantime, the little mammals survived in the undergrowth, trying not to get stepped on. As the Mesozoic wore on, they evolved their own variations and innovations. The earliest mammals still laid eggs, and a few of their descendants still do. But something new came with the appearance of marsupial mammals, for they give birth to live young who then crawl into the mother's pouch for their early development. And finally came the mammals with a placenta, live young, and no pouch. Placentals like us are now dominant except in Australia, a refuge for marsupials where placentals could not reach them until quite recently.

In addition to our mode of reproduction, two other major features of our bodies remember the early history of the mammals. The first is the active metabolism that keeps our core temperature constant at about 98.6° Fahrenheit, requiring a lot of energy and partly accounting for the tons of food that we consume in a long lifetime. The second is hair, which is a feature of mammals and no other animals.

Mammals were present through much or all of the Mesozoic—most of the Triassic, and all of the Jurassic and Cretaceous, as were the dinosaurs. In some ways the dinosaurs dominated the land fauna, at least in terms of size and variety. But the mammals must have been much more abundant, as is generally the case with smaller animals. It must have looked as if the dinosaurs had a permanent lock on the large-animal niches. That was not to be the case, however, for the dinosaurs are gone, except for their bird descendants, and in our world most of the large animals are mammals. What happened?

RELEASE! (PALEOGENE)

For a long time no one knew why the dinosaurs disappeared. Here is a wonderfully written nonexplanation from 1886: "A higher type is now standing at the threshold of being. A knell is sounding the funeral of the reptilian dynasty. The saurian hordes shrink away before the approach of a superior being. After a splendid reign, the dynasty of reptiles crumbles to the ground, and we know it only from the history written in its ruins."[17] At the time this was written, the big, obvious bones of dinosaurs were much better known than the little, inconspicuous mammal bones. But beginning in the mid-twentieth century, careful study has shown that small mammals coexisted with large dinosaurs for at least 130 million years.[18]

It seems clear that the mammal was not a "superior being." It took a kind of *deus ex machina* to get rid of the dinosaurs, and the giant impact on the Yucatán Peninsula, 66 million years ago, did the trick. The evidence is now overwhelming that the impact occurred at exactly the time of extinction of many groups of plants and animals[19] and would have caused massive environmental disturbances.[20] The environmental disruptions caused by the impact may or may not have been assisted in causing the extinction by enormous volcanic eruptions in India that had started well before the extinction event and continued afterward—this is a matter of current research and controversy.[21]

With the dinosaurs gone, except for their bird descendants, wonderful new possibilities opened up for the mammal survivors and their descendants. Most dramatic is an increase in size. Paleontologists have long recognized that mammals got much bigger just after the dinosaur extinction at the end of the Cretaceous. This event shows up clearly in statistical analyses of fossil sizes. John Alroy has shown that the average mammal in the late Cretaceous of North America, which has an excellent fossil record, weighed only about 50 grams (less than 2 ounces); within a few million years after the great extinction mammals shot up to around 500 grams (a pound) and then gradually increased to about 3,000 grams (more than 6 pounds) before dipping slightly in the last 10 million years.[22] It is a most striking pattern, strongly suggesting that the disappearance of the dinosaurs allowed the mammals to become much bigger, and this lies behind our own species' large body size.

Alroy also found that there was a similar, dramatic jump in the diversity of North American fossil mammals. At the end of the Cretaceous only about 20 species are known, but immediately after the extinction there is a steep climb to about 50 species and then a more

gradual rise to around 100 species in recent times.[23] Diagrams showing the history of the various orders of mammals make the point
even more clearly.[24] Within 10 million years after the extinction,
most of the familiar kinds of mammals have entered the fossil record,
including rodents, bats, the ungulates that would eventually lead to
horses, the carnivores, the ancestors of elephants, the whales, and, of
course, the primates.

Some of the features of our bodies reflect our membership in
the primates. Thus, in the head, our large skull and brain carry to
extremes a feature of primates in general, and our forward-facing
eyes, allowing stereoscopic depth perception, are also a characteristic of primates. The nose also tells about our primate heritage, for
unlike most other mammals, smell is a less important sense for us
than vision. Nose character is also diagnostic, for we humans are dry-
nosed primates (not wet-nosed, unless we have a bad cold). Among the
former, we are straight-nosed primates with downward-facing nostrils, as opposed to flat-nosed ones with sideward-pointing nostrils.

Another characteristically primate feature of our bodies is the
hand, with dexterous fingers and opposable thumbs, allowing us to
hold and manipulate objects, and with delicate fingernails rather than
sharp claws. Their amazing capabilities are obvious when compared
with the forepaws of a dog or cat! These wonderful hands apparently
originated for grasping to aid in a life lived in trees but, as so commonly occurs in evolution, a feature that originally served one purpose was later adapted to serve another; thus, we are able to tie knots,
to draw and paint, and to play musical instruments.

WALKING TALL, THINKING SMART, SPEAKING OUT (PLIOCENE, QUATERNARY)

What is there in our bodies that marks us as distinctly human? The
two obvious features are the upright posture that allows us to walk
on just two of our limbs, leaving the other two, with their sensitive
hands, free to do interesting things, and the large brain that lets us

think up interesting things to do. For a long time scientists debated which came first—walking upright or the large brain—but spectacular fossil discoveries in the last 40 years or so have answered the question very clearly: We walked tall before our brains got big.

That is the significance of the famous fossil called Lucy, or officially *Australopithecus afarensis*, dating from 3.2 million years ago. Found in Ethiopia in 1973–1974 by a team led by Donald Johanson, Lucy's bones make it clear that she walked more or less upright, while her skull only had room for a pretty small brain, comparable to that of a chimpanzee.

Another spectacular fossil find is Ardi, or *Ardipithecus ramidus*, also found in Ethiopia, in the early 1990s, by a team led by Tim White, Berhane Asfaw, and Giday WoldeGabriel.[25] Ardi dates from 4.4 million years ago, more than a million years before Lucy, and confirms that walking upright came before the large brain. Like Lucy's brain, Ardi's brain was of chimpanzee size, but Ardi had a strange big toe, splayed out and perhaps prehensile, suggesting that it was intermediate between toes useful for climbing trees and toes useful for walking.

So when looking at your body, you can recognize your feet and the gluteus maximus that keeps your posture vertical as slightly more ancient features than your large brain.

One final part of the body worth thinking about is the tongue. We know that it is an ancient feature, not because of fossil evidence, for soft parts rarely preserve, but because of its widespread occurrence in living mammals and vertebrates in general. The basic mammalian uses of the tongue are for moving food around as we chew it and as the organ of taste. But evolution has given the tongue additional different uses in different animals. Think of the anteater using its long, thin tongue covered with tiny hooks to extract delicacies from insect nests, or the lamprey rasping away at its victim, or the cat using its tongue to clean its fur, or the dog panting with a wet tongue to cool its body.

In the fairly recent past, the tongue has become a critical part of our ability to communicate with spoken language. Just when language developed is still a major debate in paleoanthropology because there is so little evidence to go on. But "tongue" has become synonymous with language, and to feel it in action, try reading this paragraph aloud, both slow and fast, and pay close attention to your amazing tongue, flickering and dancing in your mouth. Take a moment to recognize how it works with your jaw, your lips, and your vocal cords to make the very precise sounds that communicate the thoughts, ideas, questions, and commands that have made human history so interesting.

Reviewing the history and ancestry of our human bodies once again brings us face to face with the improbability of the human situation. What if bilateral symmetry had never appeared? What if the movable jaw had not evolved? What if the dinosaurs had not been killed off? What if other biological inventions we can barely imagine had shaped the path of evolution? As with so much else in Big History, it was a very particular and unlikely sequence of events that gave us the characteristics of our human bodies. People with all the characteristics we have just looked at first appeared in Africa, and in that sense, we are all Africans. But today we are found all over the world. How we spread so widely is the topic of the next chapter.

HUMANITY

CHAPTER EIGHT

THE GREAT JOURNEY

A SLOW-MOVING BUT UBIQUITOUS SPECIES

OUR HUMAN BODIES give us the ability to move around. We can walk, run, and swim but can do none of them very fast. Our two-legged walking is rather clumsy, and many animals can run much faster than we can. We swim, but rather slowly, and always with the danger of drowning, and our natural bodies are quite incapable of flying.

It may thus be surprising that we occupy pretty much all the habitable places on Earth. How did we get there? In the twenty-first century we can go to an airport and fly almost anywhere on the planet in a few hours, and except for remote and difficult wilderness areas, there will always be people living there. Very few other species have such a broad range of distribution.

The present chapter explores this remarkable aspect of the human situation and how it ties into the deeper history of our planet. We can start by asking what the explorers found, a few hundred years ago, as they gradually reconnected the people living all over Earth.

WAS THE REST OF THE WORLD UNINHABITED?

During the millennium before the European voyages of exploration, there were always contacts throughout the great continental mass of Africa, Asia, and Europe. Merchants traveled the Silk Road across Asia and journeyed through the Sahara, linking northern and southern Africa. A few intrepid travelers made journeys that still leave us in awe, like Marco Polo in the thirteenth century and ibn Battuta a century later. The extensive but fragile connections across these continents have been called the Old World Web by the son-and-father team of historians John and William McNeill.[1] Though tenuous linkages stretched across Asia, Europe, and Africa, people in any one place might only know about the neighboring areas and were completely unaware of a less developed American Web and an incipient Pacific Web.

For me, the breaking out from the limits of the Old World Web makes a good marker for the transition from the European Middle Ages to the modern world. Let's see what the explorers found as they emerged from Old World isolation. Was anybody there?

The first explorations by Europeans led to the discovery of the North Atlantic islands—Iceland in the 870s, and the enormous, mostly ice-covered island of Greenland in the 980s, both discovered by the Norsemen. Those Vikings could not have known that until 60 or 70 million years ago, Greenland had been attached to their home country of Norway until continental breakup and seafloor spreading carried it away toward the west. After settling in Iceland, they saw new land forming during volcanic eruptions but probably could not have conceived that the entire island was new land built by volcanism where seafloor spreading was generating new ocean. Still less could they have imagined that the huge volcanic outpourings of Iceland are fed by a portion of Earth's mantle that is abnormally hot and slowly rises up from very deep in the planet—what geologists now call a mantle plume.

The Norse explorers found no people in either Iceland or Greenland, although far away to the north there were Paleo-Eskimo people living in polar Greenland. And when the Norsemen briefly reached the part of North America they called Vinland, people were living there also.

After a 400-year interruption, the seamen of tiny Portugal in the fifteenth century began the explorations that would eventually tie the entire globe together into the single web we know today, and early in their efforts they came across the four low-latitude North Atlantic island groups. Like Iceland, these volcanic archipelagos are the products of mantle plumes, although much smaller ones. The Canaries, only 60 miles from the African mainland, had been known since antiquity and were inhabited, but nobody was living on any of the others farther out in the ocean—the Azores, discovered around 1340, Madeira around 1420, and the Cape Verde Islands around 1450.[2] It must have seemed that exploration would turn up many uninhabited places, but that was all about to change, and it would come to seem, instead, that there were people everywhere.

FINDING LANDS FULL OF PEOPLE

Motivated and financed by Prince Henry the Navigator, the Portuguese explorers, one after another, pushed slowly southward along the African coast in their tiny ships. They were sailing along the coast of the Sahara Desert, and the land became hotter and hotter, drier and drier, and ever more empty. Influenced by Aristotle's idea of a blazing hot, uninhabitable equatorial belt, they expected to find no one at all, and as they sailed along the Saharan coast, that seemed to be the case.[3] But when in 1444 they reached Cabo Verde, the green cape that marks the southern limit of the arid Sahara, they found a land full of African people. The Berbers and the Arabs who had long traded across the Sahara with camel caravans could have told them, of course, had they been on speaking terms. And as the Portuguese continued to explore onward along the African coast, they found people everywhere.

Something fascinating that we now know about the African coast where they were sailing would have astonished the Portuguese explorers, probably to the point of utter incomprehension. Although equatorial now, this part of Africa was at the South Pole during the Ordovician and Silurian, about 450 million years ago. Just imagine the amazement of the French geologists who first found Ordovician glacial deposits in the hyperarid heart of the Algerian Sahara in the early 1960s, when most geologists still believed that continents never move![4] Those glacial deposits are Earth's memory of a time when the supercontinent of Gondwana, of which Africa was a part, migrated across the frozen South Pole of the planet.

After discovering the Cape of Good Hope at the southern tip of Africa, the Portuguese continued across the Indian Ocean to Asia. Everywhere they went they found people, but that is not too surprising, for they were always within the Old World Web.

Meanwhile, in the early fifteenth century, the Chinese were undertaking truly major expeditions across the Indian Ocean to Africa and to parts of Indonesia on a scale that would never be matched by Europeans, or anyone else, ever again. The McNeills describe it like this:

> Zheng He (ca. 1371–1435), the admiral who led six Chinese armadas into the Indian Ocean between 1405 and 1433, had had such support on a scale that the Atlantic Europeans could not approach. In the summer of 1415, when Prince Henry the Navigator was taking part in a Portuguese expedition against the Moroccan town of Ceuta, Zheng He was in Hormuz, at the mouth of the Persian Gulf, on the fourth of his six expeditions. Prince Henry was about 200 miles from home, while Zheng He had sailed some 5,000 or 6,000 miles from his base. Zheng He's biggest ships were six to ten times the size of the largest Columbus later commanded, and thirty times the size of John Cabot's lone vessel of 1497. On Columbus's largest expedition (the second of his four) he had 17 ships and about 1,500 men, whereas Zheng He's first expedition counted 317 ships and about 27,000 crew.[5]

With an effort like this, it seems very strange to us today that the Chinese did not break out of the Old World Web long before the Europeans, but that did not happen. It now seems clear that the Chinese were not attempting to find new places. These voyages were never intended to be explorations of the unknown; the huge fleets were tribute-gathering expeditions intended to demonstrate Chinese dominance over peoples already known.[6] In a development that at first seems incomprehensible to us, the Chinese government terminated the series of expeditions in 1433, after Zheng He's voyages, forbidding further oceanic journeys and even the construction of oceangoing vessels. China turned inward and, as a result, it was the Portuguese who finally made transoceanic contact with China. The reasons for this termination are still being debated, but for our present purpose it is important to note that the Chinese voyagers, concentrating on Indonesia and the Indian Ocean, did not find any sizable uninhabited places because they remained within the Old World Web. When the Chinese voluntarily retired from the field, exploration was left to the Europeans.

The first real breakout from the Old World Web came when Columbus reached the island of San Salvador in the Bahamas on October 12 in 1492. Columbus himself never stopped believing he had reached India, but his contemporaries soon realized that this was the gateway to a whole New World. With the theory of continental drift and plate tectonics, geologists now understand that the little archipelago of the Bahamas, between Cuba and the Atlantic, is actually a sliver of West Africa, torn off and carried westward by the spreading Atlantic Ocean. So at the end of the fifteenth century, Columbus reached a place that had once lain where the Portuguese explorers were sailing in the middle of that century.

I imagine that Columbus was not surprised to find these islands inhabited by people, for he never gave up the belief that he had found Asia. But to us now, knowing that he had glimpsed the beginning of two continents unknown to the Old World, it is a remarkable fact that the New World was home to people who had lived there for a very long time. As European exploration of the New World continued, it became clear that both American continents were fully inhabited, from Alaska and Newfoundland to Tierra del Fuego. And yet the new continents Columbus found had striking differences from the one he came from. There were plants and animals different from any known in Europe or Africa, and the people, although definitely belonging to the species *Homo sapiens* because they could interbreed with the Europeans, had minor but noticeable differences. By the end of the chapter we will be in a position to understand how all of these differences came to be.

While Columbus was discovering the Bahamas, or shortly afterward, the Portuguese were making a wide detour westward to use favorable winds on the trip home from southern Africa and bumped into the eastward bulge of Brazil. Although they could not have known it, the Portuguese in Brazil were exploring what had once been a continuation of their new discoveries in West Africa. It was the snug fit of Brazil into the bight of West Africa that first suggested continental drift to the sixteenth-century Flemish cartographer Abraham Ortelius (only a century after Columbus's discovery) and later to the nineteenth-century French geographer Antonio Snider-Pellegrini and the early-twentieth-century German meteorologist Alfred Wegener. It was the extreme precision of the fit, demonstrated by early computer mapping, that allowed the British geologists Edward Bullard, James Everett, and Alan Smith to confirm that continental drift really had taken place.[7]

The remote places and the most remote place

Around 1600 the next major place to be discovered by Europeans was the remote continent of Australia. Like the Americas, it had always been outside the Old World Web. Australia has intriguing geological analogies to India. Both broke away from the rest of Gondwana about 100 million years ago and headed north. India moved much faster and was aimed at the southern coast of Asia, so it collided with Asia about

50 million years ago, pushing up the Himalayan Mountains and thus becoming an integral part of Eurasia, so that *Homo sapiens* have lived there since shortly after leaving Africa.

Australia moved north at a more sedate pace, and its track took it beyond the eastern end of Asia. Australia collided with the islands of the Indonesian archipelago in the recent geologic past. When the Europeans arrived, they found it fully inhabited by the Aboriginal peoples. Australia today is connected to Asia by the stepping-stones of the Indonesian islands with only short passages of water separating them. Finding people there was not truly surprising in retrospect, although the early immigrants to Australia must have been able to cross at least narrow stretches of open water.

Far to the east of Australia, across 1,000 miles of open ocean, are the truly isolated islands of New Zealand. And when European explorers stumbled onto New Zealand in the mid-seventeenth century, they found it also inhabited by the Maoris. Was there no place on Earth that people had not already reached and settled long before?[8]

Perhaps the most astonishing of these discoveries came last, in the eighteenth century, when European seafarers finally began to stumble on the extremely remote islands lost in the middle of the Pacific Ocean. As in the case of the Atlantic islands, most of these are volcanoes fed by mantle plumes. And despite the incredible distances of trackless ocean that had to be crossed to reach these totally isolated places, most of the inhabitable islands were already inhabited—even the extremely remote Hawaiian Islands. Even today we are amazed by the achievements in navigation that allowed the Polynesians to cross the vast and trackless Pacific and to reach and settle these totally isolated places.

It was not until the discovery and exploration of Antarctica in the nineteenth and twentieth centuries that Europeans, with great difficulty, finally reached another place like Iceland that was uninhabited and had never been inhabited by people. Antarctica is a fragment of the supercontinent of Gondwana that has stayed close to the South Pole for more than 350 million years, while other fragments, like Africa, India, and Australia, migrated northward to milder conditions. Perhaps the lack of human inhabitants is not surprising, for the climate is so implacably hostile and the lack of resources so complete that even today it requires major logistical feats to allow even a few people to live and work temporarily in this ice-bound wilderness.

In this brief account of the explorations that began the globalization of the world, we have seen a long string of discoveries of remote places that were found by the Europeans to be fully inhabited by unfamiliar people, bookended by the early and late discoveries of the uninhabited exceptions—Iceland, southern Greenland, the smaller Atlantic islands, and finally Antarctica. The obvious question is how and when all those people got to those remote places and where they started from. How did *Homo sapiens*, unlike almost every other species, become ubiquitous?

WHERE DID THE HUMAN JOURNEY BEGIN?

From the Middle Ages on into the nineteenth century, Christians believed that the first home of human beings was the Garden of Eden, although it was never certain where the garden was located. The theory of evolution, published in 1859, challenged that longstanding belief, but few people accepted evolution quickly or with enthusiasm. Gradually, however, the place of the origin of humanity became an interesting scientific question. In 1871, Darwin accepted the conclusion of his strong supporter, Thomas Huxley, who was arguing for an African origin because that is the present home of the apes most resembling humans.[9] Were they right?

The evidence that was needed was the oldest fossils of what are now called hominins—humans and our ancestors back to the split that led to chimpanzees. The earliest stone tools would also provide important evidence. Human fossils have always been extremely rare. Through the nineteenth century, almost all of the few ones known came from Europe, because that's where most of the people interested in such things lived and looked for them. This observational bias led some to think that humans had originated in Europe, which of course was an agreeable thought to Europeans.

Beginning in the twentieth century, however, Africa has become by far the greatest source of hominin fossils, including all such fossils older than 1.8 million years.[10] Hominin fossils of that age or younger have shown up elsewhere in the Old World—Georgia in the Caucasus, China, Spain, Indonesia, Germany—but the African finds dominate. The interpretation seems straightforward: The line leading to humans originated in Africa, and beginning about 1.8 million years ago there was at least one, and maybe more, out-of-Africa human migrations.

The fossils are supported by the presence in Africa of the earliest stone tools. Until very recently, the oldest, simplest tools known were those called the Oldowan Industry after Olduvai Gorge in Tanzania, and associated with *Homo habilis*, the first species considered to have been human. But very recently, still older tools, dating back to 3.3 million years ago, have been found at a site called Lomekwi in Kenya.[11] Younger and more sophisticated tools are found both in Africa and in Eurasia. Further support for the African origin comes from the genetic record in our DNA, considered in the next section, with the greatest genetic variance in modern humans being found in Africa. Recently it has been argued that language originated in Africa as well, based on the greatest diversity of sounds used in language also being found there.[12]

So it seems that Huxley and Darwin were right. But what can we say about how humans left Africa and spread to the rest of the world? Although the taxonomy of early humans is complicated, controversial, and based on a rather modest collection of fossils, there seems to be fair support for the view that three species of *Homo* emerged in sequence in Africa. The first, beginning around 2.5 million years ago, was *Homo habilis*, probably the maker of the crude Oldowan stone tools. Then about 1.8 million years ago came *Homo ergaster*, who produced the carefully made Acheulian hand axes probably used for butchering large prey and currently crafted by only a few hominins, like Kathy Schick and Nick Toth, as we saw in Chapter 3. Finally, about 200,000 years ago, came *Homo sapiens*, who developed sophisticated culture and a wide range of advanced tools made from stone and other materials.

What complicates matters is that although *Homo habilis* stayed in Africa, some members of both *Homo ergaster* and *Homo sapiens* migrated out of the home continent. Thus anthropologists recognize two different human out-of-Africa migrations—although, of course, each one may have involved independent departures from the continent by different groups of people.

The first out-of-Africa migration, involving *Homo ergaster*, took place much earlier, probably around 1.8 to 1.7 million years ago, based on fossils found in Dmanisi in the Caucasus by a team led by the Georgian paleoanthropologist David Lordkipanidze.[13] The story of this first migration is murky and controversial, but it appears that some of the *Homo ergaster* people migrated eastward, reaching China and Java, and evolved into *Homo erectus* by about 1.5 million years ago.[14] Others migrated westward into Europe somewhat later, and it appears that these people evolved into a larger-brained species called *Homo heidelbergensis* and, subsequently, into *Homo neanderthalensis*—the well-known Neanderthal people. All of these relationships remain controversial, and sadly our knowledge of this first great human migration is very limited because of the rarity of fossils and the fact that DNA evidence cannot be used to study any but the most recent extinct lineages.

The second out-of-Africa migration, involving *Homo sapiens*, was the beginning of the great journey that has taken members of our species to all the places where the explorers later found them. This migration came much more recently, probably beginning about 60,000 years ago, so that by the time our species appeared in Eurasia, *Homo ergaster* and its descendants had been living there for more than a million and a half years. The effects of these migrations can be seen in the record of tools. In Italy, for example, at the eastern foot of the Apennine Mountains near Ancona, the hand axes of *Homo ergaster* turn up in gravels that washed out of the mountains during the interglacial episode predating the current one; Neanderthal

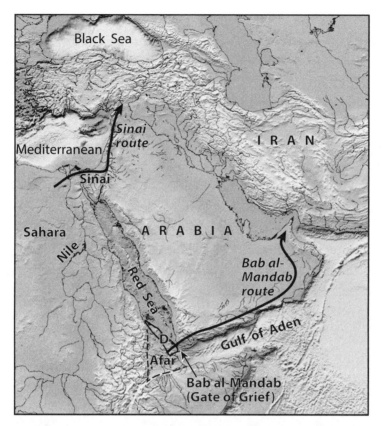

Fig. 8-1. Two possible routes by which early humans may have left Africa. "D" marks the Danakil Block of African crust that rotated counterclockwise and provided a bridge between Africa and Arabia; the dashed line shows the coastline that would have kept early humans from crossing the Red Sea–Gulf of Aden, if the Danakil Block had not rotated.

tools are found in deep valleys eroded during the last glacial advance, where their makers may have sheltered from the frigid winds; and the advanced tools of *Homo sapiens* show that their makers again spread over the uplands during the mild climate after the end of the last glacial.[15]

But how were the adventurous early *Homo sapiens* able to leave Africa? A glance at a map shows the problem. The continent is entirely surrounded by the sea, except for the northeastern corner, where the exit route leads through the Sahara and Sinai Deserts in which freshwater is extremely difficult to find. Current thinking is that out-of-Africa migrations may have taken place at either end of the Red Sea.[16]

Geological history can help explain how *H. sapiens* may have escaped the African corral through either of these routes. The Sahara-Sinai route today looks extremely forbidding because of the hyperaridity of the desert, but we know that there were rainy intervals called pluvials, quite recently in human history, for there are magnificent frescos and rock engravings on Saharan rock faces showing people and animals that today live in wetter parts of Africa.[17] The Sahara Pump hypothesis suggests that people were drawn northward into the Sahara during its wet phases and then, during the dry times, were expelled both southward and northward, with the latter groups moving on into Eurasia.[18]

The other route is at the corner where the Red Sea turns to become the Gulf of Aden. Both bodies of water formed by seafloor

Fig. 8-2. Rock engraving from the Wadi Mathendush in the hyperarid heart of the Sahara in southwestern Libya.

spreading, beginning about 25 million years ago, as Arabia broke away from Africa. Now if you simply look at the coastlines, it does not look like Africa and Arabia would fit together, but that is because of the rotation of the Danakil Block that twisted counterclockwise during the separation of Arabia from Africa. The southeastern end of the Danakil Block moved with Arabia while the northwestern end stayed attached to Africa. The low-lying desert area called the Afar is thus geologically part of the Red Sea and Gulf of Aden but was not flooded. The Danakil and the Afar thus provided a geological bridge between Africa and Arabia, and humans may have been able to leave Africa using that bridge. The Afar is also of great relevance for understanding human history, for it was there that two of the most important human fossils were found—Ardi, dating from 4.4 million years ago, and Lucy, from 3.2 million years ago, as noted in the previous chapter.

The other part of the geological story of the escape from Africa concerns the narrow strait now separating the Danakil Block from Arabia. The Arabs call this strait the Bab al-Mandab—the Gate of Grief—because of the hazardous navigation there, but it seems symbolic also of the beginning of the astonishing but troubled migration of humans all over the world. Today the Gate of Grief is 12 miles wide, much less than the 100- to 200-mile width of the Red Sea and the Gulf of Aden, but still a daunting if not impossible crossing for people with perhaps only primitive rafts. But the strait would have been much narrower during glacial times, when so much ocean water was locked up in ice on Canada that sea level was a couple of hundred feet lower. And indeed the migration of 60,000 years ago took place during the most recent glacial epoch, when sea level was lower and the Gate of Grief narrower, making the crossing much easier.

The cases of the Sinai and the Bab al-Mandab escape routes from Africa are wonderful examples of how geologic history has controlled or influenced human history and are, thus, prime exhibits in the museum of Big History.

THE ITINERARY RECORDED IN OUR GENES

The next question must be about the routes of migration that took our ancestors around the world. Human fossils are much too rare to answer this question, and stone tools and other artifacts are only slightly more helpful. But in an exciting new development, geneticists have discovered that they can use information from our DNA to help us understand the pathways and timing of our worldwide migration. In the previous chapter we saw how DNA can be used to find the relationships between species—to show where each species fits into the tree of life and to get some idea of the ages of the branching points. That approach compares the whole genome, or large portions of it, between different living species.

The situation is quite different when using DNA to track the human migration because we are all the same species with only slight genetic differences between individuals. It is also complicated by the fact that each of us inherits genes from both parents through genetic recombination, so that the slight differences in our genome are constantly changing. How can you tell the differences due to mixed-up parental contributions from the differences that have accumulated during migration?[19]

Fortunately there is a solution to this problem—two solutions in fact—because there are two parts of our DNA that do not recombine during descent through the generations. One of these is the DNA of the mitochondria that entered our cells as symbionts early in the history of life and continue to transmit their DNA from one generation to another, independently of the DNA in the cell's nucleus. Mitochondrial DNA is passed down only through the mother, so each child will have exactly the same mitochondrial DNA as the mother does, unless there is some mutation during reproduction. The father contributes no DNA to the mitochondria, so mitochondrial DNA is preserved except when there is a mutation.

The other solution is to use the DNA from the Y chromosome. Only males carry the Y chromosome, so there is no contribution from the mother, and this DNA is also conserved in males from generation to generation, apart from the occasional mutation. Since women do not have a Y chromosome, they cannot directly learn the DNA configuration of their male ancestors, but they can find this out if they have a brother or a paternal uncle.

So we have two tracers of our human ancestry—mitochondrial DNA for the female side and Y-chromosome DNA for the male line. Since these change only with the rare mutation—a marker that then propagates to all descendants—different people have slightly different records in these two kinds of DNA. None of these minor differences have caused problems that kept their bearers from reproducing successfully. As genetic anthropologist Spencer Wells has put it, "If you share a marker with someone, then you must have shared an ancestor at some point in the past."[20]

Each difference traces back to the first individual in whom the marker appeared, and by focusing on the markers in people who have lived in the same place for a long time, geneticists can reconstruct where each marker originated. It is also possible to determine the order in which the markers appeared, for older ones will be more

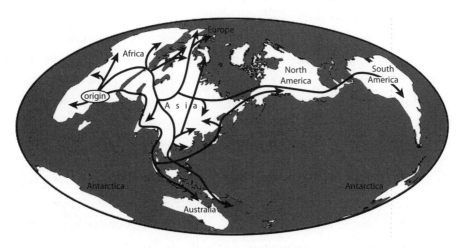

Fig. 8-3. A summary map of human migrations based on evidence from mitochondrial and Y-chromosome DNA. These migrations took place over tens of thousands of years and resulted from gradual expansion over many generations, not from intentional planning.

widespread than more recent ones. Rough dates for the appearance of the markers can then be estimated, and the final result is a map showing the routes our ancestors followed as they migrated during the great journey.

A big effort is going on now to track our human migrations using markers in mitochondrial and Y-chromosome DNA. Led by Spencer Wells, the Genographic Project relies mostly on DNA from indigenous populations to trace ancient migration pathways. It also gives nonindigenous people the chance to learn where their ancestors came from.

Although the details are fuzzy and the dates uncertain, a general picture of our ancient migration routes is beginning to emerge. It appears that one early pathway led eastward along the southern coasts of Arabia and India, through Indonesia, and into Australia, with a branch that moved north along the eastern edge of Asia and across the Bering Strait into the Americas. Another route passed through the Middle East and the heartland of Asia, with complicated branches leading to Europe, northern Asia, and then across the Bering Strait into the Americas. Various branches led to the Mediterranean, to India, to eastern Asia, and finally out across the vast expanses of the Pacific to the most remote islands.

The whole picture is too complicated for a brief description in words, but the current understanding is laid out in maps on the Web.[21] As this is very active and pioneering research, we can expect rapid improvement and even major changes in our understanding of the genetic record of the journey of humanity.

Following one theme of this book, it is worth thinking about how events in geologic history have influenced and even controlled our spread across the world. Certainly the lowered sea level during the last glacial epoch during which much of our migration took place was critical, probably facilitating the crossing of stretches of water like the Gate of Grief, the passage to Australia, and the transit of the Bering Strait by making them narrower. The rise of sea level caused by melting of the Canadian ice sheet about 10,000 years ago would have covered over the coastal routes taken by migrating human groups, making it hard now to find direct evidence of their passage.

It is also interesting to speculate about counterfactual history and ask what human migration might have been like if our species had arisen at a different point in the supercontinent cycle (Chapter 4). What if we had wandered across a huge continental mass like Pangaea with no intervening oceans? Or what if the continents had been even more dispersed than they are now, with no possibility to get out of Africa, say, and reach any other continent before the age of sailing ships, so that vast unoccupied lands would have been found by technically advanced civilizations? There are, of course, no real answers to counterfactual questions, but they do make us realize how much human history has been influenced by the world that geology has given us.

How have we traveled on our journey?

It is easy to imagine that somehow the human migration was done on purpose, as today we would undertake a journey with a destination in mind. But, of course, the humans who populated Earth did not know where they were going, because no one had been there before. And in fact, an influential book by archaeologist Albert Ammerman and population geneticist Luigi Luca Cavalli-Sforza demonstrated that there did not even need to be an intent to start a migration.[22] Simply expanding slightly in a favorable direction, generation after generation, can explain how we got to almost every habitable place on the planet.

Thus, from perhaps 5 million years ago to about 5,000 years ago, human beings traveled by walking. The astonishing migrations that took us over almost the whole globe were carried out almost entirely on foot. Everything people took with them—infants, stone tools, weapons, food, clothing, shelter, and eventually fire—had to be carried. Somehow this makes the achievement of populating the entire world in the last 60,000 years even more impressive.

Probably the first human mode of transportation to supplement walking was voyaging on rafts or in boats. Although there is no archaeological evidence for very early rafts or boats, they seem to be required by the paths of migration of *Homo sapiens* inferred from DNA studies, which involved crossing the Gate of Grief and the water between the Indonesian islands and Australia. We should not picture people inventing boats in order to make this crossing but rather utilizing craft that had gradually been developed for fishing. Recent archaeological research is showing that people reached Mediterranean islands like Cyprus much earlier than previously thought.[23]

Building better and better boats and ships has been an ongoing trend in human history, varying greatly in different parts of the world. This advancing technology has produced a wide variety of craft, from log canoes, rafts, and kayaks to oared galleys to small sailboats and huge sail-driven ships in China and Europe, to the great multihulled and outrigger canoes used by the Polynesians to cross the Pacific, to steel ships driven by steam and diesel engines, to nuclear submarines and gigantic supertankers. Small boats were surely involved in some of the prehistoric migrations, and after the reconnection of the world that followed the voyages of the explorers, large sailing ships and then steamships made possible the great recent migrations to the New World and Australia.

The next great advance in transportation came with the domestication of the horse, which greatly speeded up our pace of life.[24] Paleolithic hunters used wild horses as a source of meat, and horse bones are abundant in debris from campsites of the Solutrean culture in southern Europe about 20,000 years ago at the peak of the last glaciation. It is not yet clear when and where domestication of the horse first took place, but it was most probably on the Eurasian steppe.[25] A strong candidate is Kazakhstan, in the middle of the fourth millennium BC, with the use of the horse for transport and to supply milk "associated with the spread of Indo-European languages and culture, bronze metallurgy, and specialized forms of warfare."[26] Horses became our main mode of rapid or long-distance land transportation for thousands of years, continuing until well into the twentieth century.

Great advances in transportation came with the Industrial Revolution and the building of canals like the Erie and the construction of the railroads that made possible the great westward migration that peopled the American West. In the twentieth century we took to the air and then even to space. Now our planet is so interconnected that it is hard even to imagine the compartmentalized world of just 600 years ago.

An insight

If we think about Columbus when he first landed on the little island he called San Salvador in 1492, we can get an interesting insight into different kinds of migrations and how they have affected our world. Columbus and his men would have been struck by the strange animals and plants, by the strange people, and by the strange language and customs of those people. But how did those plants, animals, people, and customs come to be different from anything those Europeans had ever seen?

The plants and animals were different because the New World had broken away from the Old World about 180 million years ago, and the two had gradually become separated by the opening of the Atlantic Ocean. Almost 200 million years is time enough for the very slow process of evolutionary change to produce completely different descendants, isolated and protected from each other by a vast ocean.

The people looked somewhat different from Europeans but could still interbreed, so they were of the same species, *Homo sapiens*. But there were minor differences that arose biologically during the 60,000 years or so of separation, as people gradually migrated all over the planet and lost touch with each other. The animal and plant differences and the differences in people had thus arisen through biological evolution over two very different time scales.

The cultural differences were both in technology—ships versus canoes—and in language to the extent that even today linguists can barely find any commonalities between Native American and Old World languages. Those differences were not due to biological evolution but to cultural evolution, in which 60,000 years can produce enormous differences.[27]

This dramatic increase in the rate of historical change is an indication that as humans we have crossed a threshold into a new regime of Big History—the regime characterized by our large and versatile brains. In the next chapter, let us look in a Big History way at some of the achievements that make us human and at how Earth history has set the stage for these achievements.

CHAPTER NINE

BEING HUMAN

LANGUAGE, FIRE, AND TOOLS

WHAT IS IT that makes us human? As we have seen in the preceding chapter, our species lives all over the dry land of our planet, except for continental glaciers and a few other extremely inhospitable places. It is our amazing brains that have allowed us to develop the skills for living in so many environments. As the anthropologist Terrence Deacon has put it, in a memorable phrase, "Biologically, we are just another ape. Mentally, we are another phylum of organisms."[1]

This quotation brings to life the understanding that, with the appearance of humanity, we have crossed a major threshold in Big History, comparable in importance to earlier thresholds, like the production of elements in stars, the appearance of solid planets like Earth, and the emergence of life.[2] But having a big, powerful brain is not all that was needed in order for human beings to cross a major historical threshold. Imagine a supercomputer with input, perhaps, but no output devices and having no way to communicate with other computers. It might make amazing calculations, but they would stay inside and no one else would ever know about them.

However, like computers with output devices and linked up on the Web, we have the abilities to do things with our hands and to communicate using complex, symbolic languages that go far, far beyond what any other species can do with their brains. No single human being, starting from scratch with no help, could conceive, build, and use an interplanetary spacecraft or even a simple screwdriver. But our species can accomplish such a feat because we can talk about it, write down what we have learned, and build on what previous generations have done. Language has made possible what David Christian has called "collective learning."[3] As Isaac Newton most famously said, we are standing on the shoulders of giants.

As Big Historians, we would love to understand the origin of language and its early history, prior to the appearance a few thousand years ago of the tongues ancestral to the great language families of today—families like Indo-European, Niger-Congo, Papuan, Amerindian, and Afro-Asiatic. Sadly this is exceptionally difficult, and it has even been called "the hardest problem in science."[4]

The difficulty is twofold: First, unlike most of the history of Earth and life that we have looked at in this book, languages before the invention of writing leave no physical traces. Second, although we do have evidence for ancestral languages from similarities between the words in their descendant languages that have allowed the reconstruction of extinct languages like Proto-Indo-European, reconstructed languages are not easily datable, and this approach does not work at all back beyond a few thousand years ago. The problem is that languages mutate extremely rapidly—think about the difficulty today's English speakers have in understanding the Middle English of Chaucer, only 600 years ago, and our utter incomprehension of the Old English of Beowulf, just a few centuries before that. As a result of this rapid change in languages, apparent word similarities between possibly related languages are not reliable back beyond a few thousand years ago. Dating the origin of language is extremely difficult; it probably predates the remarkable artistic works of 40,000 years ago and is probably no older than the earliest stone tools of 3.3 million years ago—a really enormous uncertainty![5]

Our desire to know when language began and what tongues our ancestors were speaking as they journeyed out of Africa and around the globe looks unlikely to be satisfied in the near future, if ever. And this is ironic, for since the invention of writing, language has become by far the most important of all the ways we know about human history to the point where many people, including many academic humanists, think that the *only* history is written history.

The history of languages before writing is obscure[6] and that of written languages is overwhelmingly abundant.[7] Yet there are two other uniquely human features whose origins we can trace, which differentiate us from any other species of organisms on this planet—the use of fire and the making of tools.

THE HISTORY OF FIRE ON EARTH

The use of fire is not often on the list of critical human attributes, but when we look at what makes us human, controlled fire use might even be the most defining characteristic of our species. After all, whales may have something like a language, and chimps may use sticks as primitive tools to help them gather food. But it appears that all existing groups of human beings use fire, and no other species does—at least not in an active, premeditated way.[8]

So let us think historically and explore the use of fire by human beings.[9] A good place to start would be to ask when fire first appeared on Earth. When did it become possible for anything to burn on our planet?

Today fires of many kinds burn all over Earth, both fires planned and set by human beings and natural fires that are not our doing. So when we think about what fire is and the conditions under which it can occur, it is a surprise to realize that about 90% of Earth's history had to pass before it was possible for anything at all to burn on our planet!

That is not to say that there were no great sources of heat on the early Earth. Volcanoes have long poured out molten lava, lightning strikes locally melt the rocks they hit, and impacting asteroids and comets have caused melting and even vaporization of rocks, ever since the beginning of Earth history. Geologists who study the evidence for the history of the early Earth even speak of "magma oceans" after giant impacts, and as we saw in Chapter 2, the best current explanation for the existence of the Moon is that it coalesced from magma melted and thrown into orbit when Earth was largely or entirely melted by impact of an object the size of Mars. But none of these occurrences of molten rock resulted from a fire.

A fire, instead, is where some kind of fuel combines rapidly with oxygen, giving off heat in the process. There are different kinds of fuel that can oxidize rapidly to make fires—wood, charcoal, peat, coal, natural gas, oil, and hydrogen. We now use so much of these fuels that a major problem facing humanity is how and where to get all the fuel our civilization needs for the different kinds of fires we depend on. A second problem is how to deal with the climate change that the resulting CO_2 is starting to cause. So the historical question about when fires first became possible on Earth really boils down to when both oxygen and fuel became available.

First, let's look at the history of oxygen.[10] Free oxygen was not present in the atmosphere of the early Earth, and in fact, Earth is the only planet in our solar system that has more than trace amounts of atmospheric oxygen even now. At first this seems surprising, because oxygen is a very abundant element in Earth as a whole. In fact, as we saw in Chapter 3, the big four elements for Earth are magnesium, silicon, iron, and oxygen. But virtually all of Earth's oxygen is tied up in the solid minerals of Earth's crust and mantle, because oxygen holds the other three of the big four elements together. You can see this in the formula of olivine, one of the dominant minerals of Earth's mantle: $(Mg,Fe)_2SiO_4$. Here are all four of the dominant elements with the oxygen locked up in a solid and not available to be a gas in the atmosphere.

Oxygen is also a major part of water, H_2O, and at least water could flow around on the surface of the early Earth and evaporate to make clouds. But all that oxygen was stuck in water molecules, bonded to hydrogen, until cyanobacteria evolved the ability to photosynthesize their own organic matter from carbon dioxide (CO_2) and water, giving off free oxygen (O_2) as a waste product. Dating the origin of photosynthesis has proven very difficult, but it was probably in action by 2,500 million years ago, at the beginning of what geologists call the Proterozoic Eon and perhaps considerably earlier.

So with the appearance of photosynthesis giving off free oxygen as a by-product, the oxygen content of the atmosphere would have quickly risen to present levels, right? Not so fast! As we saw in Chapter 7, it was necessary for the photosynthetic oxygen to "rust" Earth's surface—to convert most of the reduced iron at the surface into oxidized iron—before substantial amounts of oxygen could accumulate in the atmosphere. It probably took about 1,500 million years for most of the reduced iron to be oxidized. Today most of the iron we use for industrial purposes comes from huge deposits of red, oxidized iron, called banded iron formations, which formed during that protracted period of global rusting.

Finally, sometime after 1,000 million years ago, the rusting was essentially complete and the oxygen level of the atmosphere began to rise. Now there could be fires, right? No—still not yet, because there was nothing to burn, so Earth needed to start making fuels.

Most organic matter, produced originally by photosynthesis, can burn if it is dry, but for a long time virtually all life on Earth lived in the ocean where it could not burn. It was only when plants evolved the ability to live on dry land that fires became possible. The first land plants appeared about 425 million years ago. Plants in the ocean, like seaweed, can use buoyancy for support, but land plants need to grow strong parts that can support them. This is the function of wood, which in turn made possible the first real fires on Earth. So about 90% of Earth history, from the beginning of our planet about 4,500 million years ago until the Silurian, less than 445 million years ago, had to pass before there was any possibility of fire. Once that possibility existed, lightning strikes and volcanic eruptions would have ignited fires. Rusting of iron originally kept atmospheric oxygen very low. Fire continues to limit the amount of oxygen in the air, because higher oxygen levels would make for more fires, bringing the oxygen level back down to its present stable level of about 20% of the atmosphere.

Thus land plants and land animals have co-evolved with fire for almost 500 million years of life history. Fire is both a hazard and a blessing, even to the point where some Mediterranean and Californian plants, adapted to frequent wildfires, cannot germinate without the intervention of fire.

FIRE AND EARLY HUMANS

Today wildfires are a threat to human beings because we frequently build permanent structures of flammable materials in places where wildfires are common. Fires in urban and rural areas cause huge annual losses in our modern world. It is therefore interesting to realize that wildfires may have been more beneficial than hazardous to our early human ancestors. Of course, a fire would have been a disaster to any group of hunter-gatherers trapped in its path. But think about the benefits of a fire to a group that was close to a fire but not in it. . . .

Having no permanent structures, a hunter-gatherer band would not face the loss of their treasured possessions, as we do. Coming upon a fire in progress, a band would have found animals trying to escape, making them easy prey for the hunters. And animals that did not escape the flames would provide a first taste of cooked meat, which is tastier, easier to digest, and lasts longer than raw meat. As the fire died down, there would have been burning embers that would keep the people warm on cold nights and could be nursed and tended and kept going for long periods of time, providing light and warmth and protection from wild animals in the darkness. The fire would also recycle nutrients from vegetation into the soil, making it more fertile, so that edible wild plants would grow and later make agriculture more productive.

Students of the history of fire usage distinguish between passive and active use of fire. Passive fire use was when humans came across a fire set by natural causes and took advantage of it, perhaps until it went out naturally or perhaps keeping the flames alive as long as possible by feeding the fire with twigs and protecting it. There was an evocative 1981 film called *Quest for Fire*, that dramatized the search for a new fire by a band of hunter-gatherers when their carefully tended flame is extinguished during a raid by an enemy group.

Active fire use came about when people learned how to make fire at will, using a twirled stick or striking a spark. In the tens of thousands of years since that epochal event, we have invented a great multitude of ways to use fire. It is interesting to think of all the experiments people have done, intentional or unintentional, that have taught us how to make use of fire and how to avoid all the dangers with which it threatens us. It is sobering to realize that each one of us alive today is descended from an enormous number of ancestors who played with fire but did not die from fire before they were able to reproduce. There must have been many, many who did not survive the long process of learning to use fire effectively and safely. Perhaps it is not surprising that parents have to warn children not to play with fire; that kind of behavior may have been built into us in the long millennia it took us to master the use of fire.

It is not surprising that such a protracted and fundamental development as the mastery of fire generated legends. One is a Native American legend telling that a white crow carried a burning brand from a forest fire and gave it to the people and was so blackened by the smoke that crows have been black ever since. Another is the Greek legend of Prometheus stealing the fire of the gods to give to humans.

The growing virtuosity of human fire use

From our perspective as people who may live a century at most, the beginning of human fire use, perhaps a half million years ago, seems very remote in time. But from the geological point of view, where 1 million years is the basic unit of time, it is more like just yesterday. From this perspective, it is astonishing what virtuosity of fire control we have achieved in that very short span of time!

Most of that half million years was needed just to learn the basics— how to preserve existing flames and light new fires, how to stay warm, how to do basic cooking, and how not to get burned up in the process.

Just think how many early humans must have perished in flames as those arts were slowly being mastered. Even today fires are fatal to many people each year. It has been a long and difficult task to make fire into our servant, and perhaps it is not strange that only our species has been able to do it.

With the basics mastered, people have found an amazing variety of other uses for fire in the broad sense of rapid oxidation of fuel. Think about cooking, for example. Perhaps people first cooked food by roasting meat or roots over an open fire, but today we can choose to boil or broil or sauté or stir-fry or bake or microwave our food as well. Food was not the only thing that responded well to cooking. People found that they could form clay into useful shapes and cook it in an oven, so it would become hard and permanently retain its useful shape. Thus we came to have pottery in all its many varieties—plates, mugs, cooking pots, lamps, water jugs, sinks, bathtubs, toilets, roofing tiles, bricks, and even the heat-resistant tiles that allow spacecraft to reenter Earth's atmosphere.

Cooking and pottery making could be done with simple wood fires. But people discovered that with hotter fires they could achieve other remarkable results. Ordinary quartz sand could be fused into glass and shaped into a huge variety of useful objects. Again think of all the things that can be made from glass—ordinary windowpanes and stained-glass windows, beads and ornaments, drinking glasses, eyeglasses, microscope lenses, bottles, electrical insulators, and today even touch-sensitive computer screens. Hotter fires also made it possible to extract metals from metal ores—first the copper and tin that made possible the Bronze Age and later, with even hotter fires, the iron that gave us the Iron Age. Today geology, mining, and metallurgy have given us ways to use virtually every metal in the periodic table of elements.[11]

The need for hotter fires led to the search for better fuels than ordinary wood. Perhaps the first improved fuel was charcoal, and it would be interesting to know just how the process of making charcoal was discovered. It seems a paradox—if you partly burn firewood under controlled conditions, you get charcoal, a fuel that burns at

Fig. 9-1. Cooking charcoal to drive off the steam, in the Italian Apennine Mountains.

temperatures higher than you could ever have gotten from the original wood. Who would ever have invented that? The secret is that in the process of cooking the wood to make charcoal, you drive off the water that is part of the wood. If you do it right, you end up with nearly pure carbon, which is what burns so hot because none of the heat is needed to drive off water. In many parts of the world, charcoal is still made in the old-fashioned way by carefully constructing a mound of wood, covering it with earth in which you poke a few holes to let in a small amount of air and let out the steam as the water is driven off, and slowly cooking it until only carbon is left.

Charcoal was only the beginning of the discovery of fuels that could supplement and augment firewood. Some places have deposits of peat, a product of dried-out swamps that was used extensively as fuel in the Middle Ages. Elsewhere are found coal deposits from swamps that have been buried deeply and have turned naturally into the pure carbon that charcoal makers produce artificially. Oil and natural gas have the great advantage that they can be stored in tanks and pumped around through pipes. And in the last few decades, people have learned to extract energy from other sources without actual combustion. Thus we have power plants that get their energy from the decay of very heavy atomic nuclei, like uranium and plutonium. And although it is proving very hard to achieve, we may eventually extract energy from nuclear fusion, the energy source of the Sun.

It is also worth thinking about the wonderful value of other products of fire—boiling water and steam! For the first human users of fire, steam was possibly a by-product of no special value. It was only much later that the real potential of steam was recognized, which proved to be a major contributing factor to the Industrial Revolution. The value of steam is that when water turns to steam, it expands powerfully. When inventors like Thomas Newcomen and James Watt figured out how to harness that power and use it to do work, human lives were forever changed.[12] No longer did all work need to be done by the muscles of people and animals. Instead the power of steam and later other energy sources, could do the work for us. The Industrial Revolution sounds like a truly terrible time for the people who lived

Fig. 9-2. Final launch of the Space Shuttle *Endeavor,* May 2011.

through it at the lower ends of the social scale. But it liberated the generations that followed from the unending physical toil that was once the unavoidable lot of most of humanity.

No longer did we need to walk everywhere, as our ancestors had to do. Great steam engines pulled trains until the mid-twentieth century, and now diesel-electric engines do that job. Tiny repeating fires in the cylinders of automobiles allow us to drive wherever we desire. Larger continuous fires in jet engines have let us realize the ancient human dream of flying. And people and instruments are now launched into space, riding on huge pillars of fire.

Tools and the Bronze Age

Advanced language and skilled fire use are two of the capabilities that have made humanity seem mentally like a whole new phylum of organisms. Let's now turn our historical mindedness to another of the distinguishing traits of human beings—the complex tools we can make and use because of our brains and our amazingly dexterous hands. We have already seen, in Chapters 3 and 8, that stone tools were central to human life for many millennia.

The first metal people learned to use was copper, which characterizes the period known as Chalcolithic, meaning stone plus copper. Ötzi, the man from about 3000 BC who was found frozen in the ice of the Alps (Chapter 5), was carrying a beautifully polished copper axe.[13] Because copper is not a strong metal, it was largely replaced by bronze when smiths learned to make that much harder, artificial metal by mixing copper and tin.

At the beginning of Homer's *Odyssey*, recounting the aftermath of the Trojan War, the goddess Athena, in disguise, visits Telemachus, son of the absent Odysseus. She tells him, "As for my arrival in Ithaca, I came with my own ship and crew across the wine-dark sea. We are bound for the foreign port of Temese with a cargo of gleaming iron, which we mean to trade for copper."[14]

Why would Athena claim to be in the odd business of trading

hard iron for soft copper? At the time, limited amounts of iron probably came from iron meteorites because there were not yet fires hot enough to refine iron ore or to work iron. As a result, iron was of less importance than copper because this was the Bronze Age, and bronze is an alloy of copper and tin. Almost as hard as iron but workable at lower temperatures than needed for shaping iron, bronze was the first metal humans could use to make tools that were easily shaped and able to hold a sharp edge.

Tools made of stone, bronze, or iron are commonly found in archaeological excavations, and archaeologists have found it useful to divide up early human history into the Stone Age, followed by the Bronze Age, and then the Iron Age. Clearly this was not the only interesting change that was going on, for there must have been developments in language, social organization, religion, and many other areas of human activity, but tools of stone, bronze and iron leave a better archaeological record.

So where was Temese, and why did Athena say she was going there? Temese is generally identified with modern Tamassos, on the island of Cyprus, in the northeastern corner of the Mediterranean. Cyprus was the greatest source of copper in the ancient Mediterranean and Near East. The name of the island even derives from "copper."

The dates for the start of the Bronze Age vary somewhat from place to place because copper and tin are not available everywhere and because the necessary technology was invented or acquired at different times in different places, gradually diffusing so that the area of bronze working gradually expanded. For the eastern Mediterranean—Crete, Greece, Turkey, Cyprus—we can think of the Bronze Age beginning gradually around 3500 to 3000 BC and ending very much more abruptly at about 1200 BC.

The ending makes for another great historical mystery because a flourishing Late Bronze Age civilization, extending all across the eastern Mediterranean, simply became extinct within the short span of a century or less. This extinction was so complete that Homer, writing 400 years later, had only legends to go on. The great palaces and cities where the Late Bronze Age civilization was centered simply vanished and were forgotten—so thoroughly forgotten that they were believed to be no more than legends until pioneering archaeologists, like the German Heinrich Schliemann, began to uncover them in the nineteenth century, about 3,000 years later.

How could a civilization simply disappear? Historians have attributed the disappearance of the Bronze Age cities to several possible causes, including drought, migrations, and the rise of iron working, but two hypotheses seem particularly interesting to me. One, proposed by geophysicist Amos Nur at Stanford, is the possibility of a widespread episode of earthquakes clustered together in time. The other, due to Robert Drews, a historian at Vanderbilt University, suggests that barbarian tribes learned to defeat the chariot-borne archers on which the Late Bronze Age cities depended for defense, enabling them to subdue and destroy one city after another.[15] Whatever the cause, the toll was horrifying. In Cyprus alone, around 1200 BC, Paleokastro was burned, Ayios Dhimitrios was abandoned, and Sinda, Kition, and Enkomi were all burned. It took several hundred years for Middle Eastern civilization to recover.

The ancient mines of Skouriotissa

Let us think again about the Bronze Age civilization and specifically about the source of the raw materials for its characteristic metal. Where did the people of those times get the copper and the tin that went into making bronze? In considering this kind of question in Chapter 3, we came to another historically minded realization: Natural resources do not just happen—each deposit of minerals or coal or oil or gas accumulated for a reason, and each has its own history. Geologists work out these resource histories as part of reconstructing the broader history of Earth, and understanding them has been critical in discovering the vast natural resources on which civilization depends.[16]

Bronze is made mostly of copper, alloyed with a small percentage of tin. It is now clear that most of the copper for the Mediterranean–Near Eastern Bronze Age came from the island of Cyprus. For a long time, however, the copper-mining history of Cyprus had been nearly forgotten, like so much else about the Bronze Age.

In the second decade of the twentieth century the geologist Charles Gunther and the father-and-son team of American mining engineers, Seeley and Harvey Mudd, rediscovered and developed the copper of Cyprus, because copper wire had become important for the commercial development of electricity. Gunther recognized ancient slag heaps at places like Skouriotissa, where Bronze Age miners had dumped the residue after copper had been extracted from copper ore, and the Mudds reopened the ancient mines using modern techniques. Cyprus eventually became a major exporter of copper until 1974, when Turkey invaded the island and the cease-fire line separating Greeks from Turks settled down close to the mines in Greek territory, which were then abandoned because they were within range of Turkish artillery.

Today you can visit Skouriotissa, as we did in 1989 with George Constantinou of the Cyprus Geological Survey and the American geologist Eldridge Moores. You can see two generations of copper mines, the first abandoned probably when the Bronze Age ended and the other abandoned about 1974. There you can recognize evidence for major historical advances in the technology of mining and metal refining. The modern mines are enormous open pits with now decaying terraces where quarry trucks carried out the copper ore until 1974. Wandering around the abandoned modern mine pits of Skouriotissa, we saw patches of ochre, a bright orange, earthy material containing about 1% copper. Ochre is familiar to artists as a pigment for making paints, and it was interesting to see where it comes from.

Techniques for refining ores are so good today that everything in the Skouriotissa pit could be ore, even ochre. Bronze Age miners were not nearly so efficient at extracting copper, so they could only use the highest-grade ore with perhaps 2% copper from seams they followed using underground tunnels. Today you can see those tunnels where they were intersected by the walls of the modern open pit mines. Archaeologists have explored those tunnels, finding ancient mining tools and occasionally the bones of an unfortunate miner.

Fig. 9-3. The copper mine at Skouriotissa in Cyprus. The earliest mines, using underground tunnels, were probably abandoned at the end of the Bronze Age about 1200 BC. The twentieth-century mine that produced this open pit was abandoned after the Turkish invasion of Cyprus in 1974.

The slag piles themselves show that refining technology was improving through the centuries of copper mining in antiquity. George told us about his analyses of the copper remaining in the slag. Lower in the piles, the slag was dumped earlier and has more copper remaining; the younger slag, higher in the pile, has less copper, because as time passed, the miners were getting better at extracting the copper from the ore.

So copper mining in ancient Cyprus has its own history, extending over many centuries. George told us about the puzzle of why it was possible to mine copper there for so long in antiquity. To extract copper from ore you need to heat it, which was done with charcoal fires that burned hotter than wood fires. The improved efficiency of extraction revealed by analyses of the copper remaining in the slag surely came from improvements in fire technology, as we saw earlier in this chapter. But the big question George raised with us is where the charcoal could have come from.

As we have seen, charcoal is made by partially burning wood under controlled conditions, enough to drive off the water in the wood and leave just carbon behind. This is a useful thing to do because pure carbon burns hotter than wood. So the ancient miners needed to cut down forests to get the wood for making charcoal. George measured the volume of slag around the ancient mines, and he said his calculations of the charcoal that would be needed to refine that much copper would require cutting down a forest at least 16 times the size of the entire island of Cyprus![17]

George said that in antiquity copper ore had been found in a number of places in the eastern Mediterranean and had been mined and refined with charcoal from the primeval forests. But in those other areas, once the original forest was cut, it would not regrow, and mining came to an end. What was different about Cyprus?

The difference was climate and rainfall. Elsewhere in the eastern Mediterranean the climate was drier. Trees remained there in early antiquity as remnants of larger forests that had flourished during the colder, wetter conditions in the last ice age that ended about 10,000 years ago. But once cut, the forests were gone for good.

Cyprus was different—it had more rainfall, so forests would grow back after cutting. On Cyprus, wood was a renewable resource. The importance and value of this resource in antiquity can be seen in documents from the Amarna tablets, found in upper Egypt. In a series of letters of the fourteenth century BC, the King of Alasiya, thought to have been Cyprus, agrees to send copper to the Pharaoh in exchange for "silver in very great quantities . . . the very best silver."[18]

Cyprus was thus the copper source for antiquity because its copper ore occurred in a wetter climate that allowed forests to regrow. And Cyprus was wetter because of its high Troodos Mountains. On some satellite images you can see the entire eastern Mediterranean baking under a cloudless sky except for clouds, and perhaps rain, over the mountainous part of Cyprus. Following the meandering investigations that make Big History so much fun, we now turn to the Troodos Mountains with Eldridge Moores as our guide, because the Troodos are enormously important to the history of how geologists came to understand the way Earth works, because Eldridge was a central figure in that revolutionary scientific development, and because it all ties into the origin of the copper for the Bronze Age.

Where the copper came from

Eldridge was one of my roommates when he was a post-doc at Princeton in the mid-1960s, and he is now a professor of geology at the University of California, Davis. Eldridge was looking for an interesting project. One night in the Geology Library at Princeton, as he was studying a geologic map of Cyprus,[19] he realized that he was looking at something that might be really important. Years later, as Eldridge was telling me about that night in the library, he said the hair was standing up on the back of his neck just remembering it.

What got Eldridge so excited was an indication that there were enormous numbers of *dikes* in Cyprus. Dikes are a common geologic feature—they form when deep rocks break apart and a crack forms that is filled with molten rock from still deeper down that then cools and solidifies. Usually a dike bears witness to a few inches of extension at most—the width of the dike. But the map of part of the Troodos Mountains and pictures in the book that came with it showed that *everything* was dikes—young dikes cutting older dikes cutting still older dikes. This was not telling about inches of extension but about tens of miles of extension! R. A. M. Wilson, the geologist who made the map, called them *sheeted dikes*, but writing in 1959, he could not possibly have realized their significance. Eldridge, however, did know about Harry Hess's 1960 suggestion that oceans form by seafloor spreading.[20] Eldridge realized that sheeted dikes are just what you would expect in an ocean crust that formed by seafloor spreading with hundreds or thousands of miles of extension! Maybe this was a map of ancient ocean crust, somehow uplifted and exposed on the island of Cyprus! Just occasionally in a life in science, you may realize that you are looking at something of fundamental importance, and this was such a hair-raising moment for Eldridge.

Eldridge and other geologists studied the rocks of the Troodos Mountains with great care, and gradually they came to understand oceanic crust and how it is generated by seafloor spreading.[21] It would have been very difficult to do this if ocean crust was only to be found far beneath the sea where it could not be reached. It was very kind of Nature to bring up a few pieces of oceanic crust in places like Cyprus where geologists can walk around on it and study it! We use the term *ophiolite* for the whole suite of rocks found in these pieces of ancient oceanic crust, now exposed at Earth's surface. The sheeted dikes are the most distinctive rocks, but there are several other components of ocean crust, including "pillow basalts" that form big lumpy masses where some of the basalt magma came up through a dike fracture and poured out on the seafloor.

One early picture from a deep dive on the Mid-Ocean Ridge, published in the *National Geographic Magazine* in November 1979, contains the key to the origin of the copper. In an area of hydrothermal vents and the black smokers of Chapter 7, it shows a white crab crawling over a surface of basalt covered with what looks like a dusting of orange powder.[22]

Orange! A very unusual color in the deep sea, but one you can see in the ophiolite rocks of the Troodos Mountains in Cyprus. It is the color of ochre. This is the color of the ancient copper mines of Skouriotissa in Cyprus! Ochre is slightly altered copper ore, and now it becomes clear—the copper ores of Cyprus that made the Bronze Age possible were formed in black smokers on an ancient seafloor, where the oceanic crust that is now the Troodos ophiolite was made by seafloor spreading. Later a bit of that ocean floor was driven up above sea level to expose the copper deposits for mining on Cyprus, and also pushing up the Troodos Mountains, whose elevation and rainfall made possible the regrowth of the forests needed to make charcoal for refining the copper ore.

This origin for the copper of Cyprus was confirmed when a Russian oceanographic team explored a place they named the Snakepit Hydrothermal Area on the crest of the Mid-Atlantic Ridge at about 23° north latitude.[23] There in a vast expanse of basalt on the seafloor they found active black smokers as well as extinct chimneys where the hot water no longer flows, and they mapped areas of metal-rich sediment and massive metal ores rich in copper. The Russian scientists were studying a presently active analogue to the hydrothermal vents that produced the copper ore of Cyprus—the ores that made possible the Mediterranean and Middle Eastern Bronze Age that was such an important phase in the development of Western civilization. The attribution of the Cyprus copper ores to mid-ocean hydrothermal vents was strengthened in 1999 when careful searching in Cyprus turned up fossils of tube worms.[24]

TIN FOR THE BRONZE AGE

We've now traced the source of the copper for the Bronze Age to Cyprus and linked it to black smokers in the deep sea. We've seen also that the source of charcoal to refine the copper ore was in the renewable forests of Cyprus, made possible by rainfall on its high Troodos Mountains. But one mystery still remains: Where did the Bronze Age smiths get the tin they needed to mix with copper to make bronze?

Both copper and tin are soft metals because their atoms can easily glide over each other, allowing the copper or tin to deform. However, when they are mixed to make the alloy called bronze, the different-sized atoms of tin interfere with the smooth deformation of copper, and vice versa. This is what makes bronze so hard, and this was one of the great discoveries that made civilization possible. Copper alone is not enough; if there is no tin, there can be no bronze.

Tin and copper are very different chemically, so the hydrothermal vent systems that harvest copper from seafloor basalt and turn it into ore bodies at black smokers do not also concentrate tin. Tin comes instead from granites, which are found in mountain belts on continents, not in the deep sea. There is no tin ore on Cyprus, so where did the Bronze Age tin come from?

The most famous source of tin in the Old World is at Cornwall in the southwestern part of England. For a long time historians supposed that tin for the Bronze Age was mined in Cornwall and shipped to the Near East to be mixed with copper from Cyprus to make bronze.

Cornwall always seemed an unlikely source of tin to me. Cornwall and Cyprus are more than 3,500 miles apart by sea, and part of that sea voyage would cross the stormy, dangerous waters of the Bay of Biscay and the English Channel. I couldn't help thinking of the Bronze Age sailor, Odysseus, whose voyage home from Troy to Ithaca, fraught with many dangers, took him ten years. Of course, that's a legend and a story, but it's a reminder that sea travel was not easy in the Bronze Age.

Thus it was gratifying to learn of a remarkable discovery in the 1980s by the Turkish-American archaeologist, Aslihan Yener, of the University of Chicago.[25] Exploring ancient sites in Anatolia, or eastern Turkey, she came across forgotten Bronze Age mines that had once produced tin. Very little tin remained, because the miners had extracted almost all of it, but excavations revealed pottery crucibles for smelting tin from ore. Chemical analyses showed tin remaining in the crucibles. It is thus clear that there once were rich tin deposits in Turkey, separated from Cyprus by only 50 miles of sea. It seems very likely that it was the juxtaposition of Cypriot copper and Anatolian tin that made the Bronze Age possible.

The tin-ore granites of Anatolia occur in a region of continental mountains, completely unrelated to the ancient oceanic crust of Cyprus. These two geologic provinces, originally very far apart, have been brought into close proximity by the plate tectonic motions that Alfred Wegener and Harry Hess first recognized. As we so often find in Big History, historical episodes in the remote geologic past have set up the conditions for much later episodes in the history of humanity, like the Bronze Age. Without the particular history Earth has had, human history would have indeed been so very different. This is yet another example of the extreme improbability of the human situation that Big History has given us.

EPILOGUE

CHAPTER TEN

WHAT WAS THE CHANCE OF ALL THIS HAPPENING?

THE CHARACTER OF HISTORY: CONTINUITY AND CONTINGENCY

HAVING NOW LOOKED at all the regimes of Big History, we might finally ask whether we can see regularities in the unfolding of the past or whether history is all disorganized confusion—just "one damn thing after another." Are there laws that control the unfolding of history? Beginning with Isaac Newton in the seventeenth century, scientists have discovered that there are unbreakable mathematical laws governing the motions of objects and all transformations of energy. This discovery suggested that beneath the seemingly capricious events of daily life there might be an underlying order and that it might even be possible to discover fundamental laws of history itself. But the search for these hoped-for laws of history has not been very successful.[1]

Planets orbiting stars and glaciers creeping down mountainsides are the kinds of things in the history of the regimes of the Cosmos and Earth that do obey the mathematical laws of physics, and their motions can be calculated. At some point in Big History, however, something different appeared, making it seem very unlikely that well-formulated laws of human history, especially mathematical ones, can be discovered or that they even exist. What changed was the appearance of life, in which each cell or each multicellular organism is an independent *agent*, competing with other agents—seeking to acquire energy and nourishment and passing its traits on to its offspring if it is successful.

The appearance of living agents took our planet beyond the realm of the phases for which physicists can discover natural laws—plasmas, gases, liquids, and solids—and brought into being matter organized in far more complex ways. This, I would suggest, was also the beginning of the regimes of Big History—Life and Humanity—for which natural laws may simply not exist. If this is correct, and there are no deterministic laws governing the history of living organisms or human beings, is there any other way to make sense out of history? Can we find, for example, regularities or patterns in the unfolding of history?

In a provocative book entitled *Time's Arrow, Time's Cycle*, Steven Jay Gould argued that there has long been a difference in views of the past between those who saw history as directional and those who considered it to be cyclical.[2] He argued that the conflict between those two views not only influenced how scholars interpreted human history but also was a central intellectual conflict among the early geologists who developed our first serious understanding of Earth.

Time's arrow versus time's cycle was a reasonable dichotomy when geologists and historians had only *words* to describe the past. The "decline of Rome" and the "rise and fall of empires" were provocative descriptions of the past, suggesting arrows and cycles. Now, however, especially in geology, we have a great richness of quantitative data sets—numerical measures of how things have changed in Earth's past.[3] In studying those plots of history I find it hard any longer to see the arrow/cycle dichotomy as fundamental.

The problem is one of time scales. If we look at the temperature history of Earth's surface, we see trends or cycles depending on the time interval we choose.[4] For the last 10,000 years, temperature has been remarkably constant, but taken over the last million years, the temperature plot is cyclical with glacial and interglacial times alternating over 100,000-year cycles. But on shorter or longer time scales, there are cooling trends and warming trends.

As I look at the way history has unfolded across all the regimes of Big History, I think I see a different dichotomy. On the one hand, I see *continuities*, made up of trends and cycles, combined in various ways at various time scales. On the other hand, there are *contingencies*—rare events that make significant changes in history that could not have been predicted very far in advance.

Contingencies are everywhere. In our personal lives, we can pass through long periods of continuity, going to work and returning home every day in a cyclical pattern, and meanwhile gradually getting older and perhaps wiser, which are trends. And then, completely unexpectedly, contingency can strike, and we can fall off a ladder, or fall in love, and nothing will ever be the same again. Human history is riddled with contingency, and this is part of what makes it impossible to find laws controlling history. Contingency is particularly dramatic in warfare, with the outcomes of battles turning on such unpredictable circumstances as the way the wind was blowing or the finding of accidentally lost orders.[5]

Although contingency surrounds us and we can often recognize it, defining contingency turns out to be quite difficult. I have not yet found a satisfactory definition or been able to construct one. My current thinking is that for an event to be considered contingent, it needs to be (1) rare, (2) unpredictable, and (3) significant. But each of those qualities involves ambiguities. And when we try to understand what causes contingency, the situation in the two inanimate regimes of Big History—Cosmos and Earth—seems to be completely different from the situation in the Life and Humanity regimes.

Let us first consider contingency in the Cosmos and Earth regimes, taking the impact that finished off the dinosaurs as an example. Then we will look at contingency in the Life and Humanity regimes, with the Spanish Armada as an example. And, finally, let's think about the astonishing contingency that lies behind the presence here on Earth of each one of us. In each case we will see to what extent the rare-unpredictable-significant criteria are applicable.

CONTINGENCY IN THE COSMOS
AND EARTH REGIMES

For the two inanimate regimes, the "rare" and "significant" aspects of contingency are not difficult to understand. In the power-law distribution of sizes for many natural objects and events, like the diameters of impacting objects and the magnitudes of earthquakes, larger objects and events are rare compared to smaller ones.[6] Larger events are significant over broader areas and touch more people.[7] It's the "unpredictable" aspect of contingency that is most interesting.

It once looked to scientists as if everything that happens in the universe is completely controlled by the deterministic, mathematical laws of Nature. These laws seemed to be fully predictive. In the early nineteenth century, the French mathematician Pierre-Simon Laplace argued that if a sufficient intellect—a super-super-computer, we might say—were given detailed information about the position and velocity of every particle at some time in the past, it could predict the future with complete accuracy. If this view were correct and applied to all the atoms that make up a human being, it would imply that we have no free will.[8]

But that was before the twentieth-century discovery of the limits to predictability. We now understand that even with inanimate materials that clearly obey the mathematical laws of Nature, in certain situations there are complications that make the future completely unpredictable. For most scientists it was a real surprise to find out that a fully deterministic system may not be predictable. Here are some examples, ordered from larger to smaller scale.

In a solar system with three or more objects, the equations of motion cannot generally be solved, although the planetary motions themselves can be traced out with computer-intensive calculations. When this is done, it turns out that tiny differences in the initial position and velocity of a planet lead to significant differences in its position and velocity after tens of millions of years. This means that over the long haul, planetary motions are unpredictable, a situation known as orbital chaos or deterministic chaos.

Fracturing of solids like rocks is unpredictable in detail, depending sensitively on microcracks and on flaws or dislocations at the atomic scale. Some tiny fractures will cascade into giant earthquakes but most do not. This is what makes it apparently impossible to predict earthquakes in a precise way—to say, for example, "there will be a magnitude such-and-such earthquake at a specific time and place"— although probabilistic estimates of the likelihood of a given size earthquake in a general region over a rough time frame are indeed made.

Turbulence in fluids is also chaotic, and this places weather, storms, and ocean currents in the category of phenomena that are unpredictable in detail.[9] Belousov-Zhabotinsky reactions are a class of chemical phenomenon in which circular or spiral patterns appear in a petri dish, growing and interacting in a chaotic, unpredictable fashion.[10]

Finally, there is the logistic equation, a very simple and easily derived mathematical formula. The logistic equation contains a squared term, and when it is applied iteratively, with the result of each calculation fed back as the starting point of the next calculation, the results are chaotic in the sense that the tiniest difference in starting condition produces completely different results down the line. Originally discovered by Robert May when he was trying to predict the number of individuals of a species over many generations, given different rates of reproduction, this has become a classic example of deterministic chaos and mathematical unpredictability.[11]

It thus seems clear that despite the rigorous mathematical laws of Nature, many aspects of our human situation, from the scale of the solar system down to the scale of a petri dish and smaller, are fundamentally unpredictable.

IMPACTS AS CONTINGENT EVENTS

Contingencies have impressed me strongly because of the work I've done on the extraterrestrial impact that caused the extinction of the dinosaurs. Let us look at the character of that event as an example of contingency in the Cosmos and Earth regimes.

We humans exist only because of the extinction of the dinosaurs. Dinosaurs were the dominant large animals on Earth from about 200 million years ago until their disappearance 66 million years ago.[12] Mammals were around for most or all of that time but never got very large or very diverse until the dinosaurian competition was removed. After that cataclysmic event, mammals began a rapid diversification and some lineages increased enormously in size, filling the large-animal niches formerly occupied by dinosaurs.

It is hard to imagine that an outside observer, dropping in to visit Earth 150 or 75 million years ago, could ever have predicted that mammals, including large and intelligent ones, would ever become the dominant animals on this planet and that dinosaurs would be represented only by their descendants, the birds, most of them quite small in size. The dominance of mammals was the result of a contingent rerouting of major pathways in life history.

The evidence now strongly indicates that the dinosaur extinction was the result of the contingent impact of an asteroid or a comet that created the Chicxulub Crater, 110 miles in diameter, now buried almost a mile down beneath younger sediments, under the surface of the Yucatán Peninsula of Mexico, as we saw in Chapter 1.[13] It is also possible that the enormous volcanic eruptions of the Deccan basalts in India at about the same time may have played some role in the extinction as well, as mentioned in Chapter 7.

The impact event meets all of my proposed criteria for contingency—impacts that size are very *rare*, the Chicxulub impact was *unpredictable* over the long term because of orbital chaos, and it was very *significant*, completely changing the course of evolution. Let us think about each of those three features of the Chicxulub impact.

First of all, impacts making craters the size of Chicxulub have been extremely *rare* since very early in Earth history. Chicxulub, 66 million years old and 110 miles in diameter, is the largest known crater that formed in the last 540 million years, during the Phanerozoic—the time for which we have good fossil control—and indeed for a much longer time than that. The only larger craters preserved on Earth are very much older—Sudbury in Canada, 160 miles in diameter and

Fig. 10-1. Artist's conception of the impact event at Chicxulub, in Mexico's Yucatán Peninsula, that caused the mass extinction event 66 million years ago, after which mammals replaced dinosaurs as the large land animals.

1,850 million years old, and Vredefort in South Africa, 200 miles in diameter and about 2,025 million years old.

When we consider the size of the solar system and the paucity of potential impactors with diameters of several miles that might make Chicxulub-scale craters, it is understandable that these have been rare events in the last 3 or 4 billion years (although common when Earth was first assembling). Most asteroids (about 99%) and a third of the far less abundant visible comets never come closer to the Sun than the orbit of Earth and, thus, have zero chance of hitting our planet.[14]

Even for Earth-crossers, the chances of hitting Earth are extremely small, and here's a way to see why. Make a circle with your left thumb and forefinger to represent the orbit of the Earth around the Sun—an imaginary dot in the center of the circle of your thumb and forefinger. Then link it with a circle of your right thumb and forefinger, representing the orbit of a killer asteroid. Now imagine the circle of your left hand changing from fingers to a filament 1/100 the diameter of a human hair, representing the path swept out by Earth. And imagine your right-hand fingers shrinking to a filament 1,000 times thinner than the left-hand one, representing the path of the asteroid. Only if those hairs pass through each other is there any chance of an impact, and since impactor orbits move around in time, there is only a chance of an impact for a short time, before the hairs no longer intersect. And even if the orbits do intersect briefly, there is only a tiny window of time, a few minutes at most, in which both impactor and Earth are at the intersection point at the same time—otherwise, no impact. For any given asteroid or comet, the chance of impact is extremely small!

If there are vast numbers of potential impactors, as there are for tiny objects the size of sand grains, then statistically there will be many impacts. This is why on most clear, dark nights you can see an occasional meteor, or shooting star, which is a sand-size meteorite burning up by friction as it enters Earth's atmosphere at high velocity. But for objects large enough to cause an extinction, statistically impacts will be extremely rare.

Was the impact *unpredictable*? Here we meet a fascinating paradox. On the one hand, nothing can be more exactly calculated than orbital motions in the solar system, which are fully determined by the laws of physics. That is why astronomers can predict precisely when and where each eclipse will be visible on Earth, centuries in advance, and why spacecraft can be sent on trajectories to meet up with planets years later.

On the other hand, there are two factors that can confound these predictions. If our potential impactor is a comet, with pockets of ice in a matrix of rocks and dirt, the heat when the comet gets close to the Sun will vaporize these ice pockets, creating little jets of steam that will deviate the comet's trajectory in small but unpredictable ways.

More interestingly, it is now understood that orbits in a system with more than two bodies, like our solar system, are inherently unpredictable over the long term. This is called deterministic chaos, and it means that even if you knew extremely well the positions and motions of all the solar system bodies early in its history and had unlimited computing power, you could not possibly calculate the positions and motions today.[15] So there is a subtle but important distinction reflected in the term *deterministic chaos*—the motions of an orbiting body are fully *determined* by the laws of physics, but they are *unpredictable* because we can never know the starting conditions well enough. Over the long term, the tiniest differences in the original positions and motions produce huge variations in today's solar system—a state called "sensitive dependence on initial conditions." In that sense, the dinosaur-killing impact was unpredictable.

When we ask whether impacts are *significant*, we run into another question of scale. It is hard to deny that an impact like Chicxulub, removing about half of the genera of animals on Earth including the large dinosaurs and bringing a group of previously small animals—the mammals—into dominance, was significant by any measure. Smaller impacts are significant for anything living at ground zero, like the meteor that burned up over Chelyabinsk in Russia on February 15, 2013, but not at the scale of the entire planet. Still smaller impacts, like the meteor streaks several times a night, must be insignificant by any measure.

CONTINGENCY IN THE LIFE AND HUMANITY REGIMES

When we leave the inanimate world of the Cosmos and Earth regimes and cross the threshold to the living organisms of the Life and Humanity regimes, contingency becomes both more pervasive and more difficult to understand. Things are now much more complicated than they were with just the four phases of the physicists because each organism is an "agent," selected by Nature to be good at obtaining nourishment and at reproducing. Each agent is competing with others of its kind to eat and to avoid being eaten. As evolution progresses, new approaches to this double task emerge, the opportunities for contingency are many, and the result is the life history that did happen.

Examples are everywhere. The evolution of eyes in predatory trilobites about 540 million years ago is thought to have triggered the evolution of hard, protective shells in many kinds of prey species, which marked the beginning of the abundant fossil record. The appearance of a new and very virulent bacterium gave rise to the Black Death that completely rerouted the history of the European Middle Ages.

As a further example, I like to think about the fact that we have two arms and two legs. Few animals other than birds walk on two legs, and it took a major evolutionary development, from *Ardipithecus* to *Australopithecus* to *Homo*, for us to be able to walk on two of our four limbs, freeing the other two with their hands, to learn to make and use tools, which is one of the key aspects of the human situation. Our four limbs are descended from the four fins of the lobe-finned fishes like *Tiktaalik* and *Acanthostega*, around 370 million years ago, which evolved into the first land animals.[16] Thinking in the counterfactual mode, suppose that those fish had had six fins. Might land animals ever since have had six limbs, making it easy for them to walk on four legs and still have two arms free to make tools? Could human-like intelligence and toolmaking have appeared much earlier with Earth's intelligent animals looking something like centaurs?

With the appearance of intelligence and language and tools that marked the beginning of the Humanity regime, contingency became even more widespread. This effect is not easily quantified, but it is interesting to think about it. For example, in living organisms other than humans, each individual *is* an experiment. If the individual's genes give it a survival and reproductive advantage, they are likely to be passed on. The time for each generation to succeed the previous one sets a limit on how fast genetic contingencies can pile up.

But our brains allow us to do thought experiments about possible futures—for example, should I be a musician or a scientist or a bank robber when I grow up? Each of us can run many such thought experiments during our lifetime, and deciding which mental experiment to carry out in real life allows enormous contingencies to direct history down any one of myriad possible paths. Think how different the twentieth century would have been if Adolf Hitler had stayed with his early career as an artist.

The neurophysiologist William Calvin has suggested that a similar process of constructing multiple scenarios and choosing between them takes place constantly in the human brain.[17] According to Calvin, even in a fast-paced conversation the brain generates many possibilities for what to say next and then chooses which one to say, perhaps on the basis of previous experience in similar situations. Brilliantly creative ideas and things one will always regret having said arise in this way, through the brain's ability to generate lots of possible scenarios. Systems that function in this way, by generating multiple possibilities and choosing among them, have been called Darwin machines.[18] Among them, both evolution and the human brain seem to have been major contributors to the enormously contingent path of life history and human history.

CONTINGENCY IN HUMAN HISTORY: THE SPANISH ARMADA

Contingency is everywhere in human history. As a geologist, coming from a field where it is much less obvious, I find contingency very interesting. David Christian, the founder of Big History, once told me that for him, coming from the humanities, contingencies are so overwhelmingly abundant that they don't interest him very much; he prefers to try to identify continuities. Just to give the flavor of contingency in human history, let's look at a case from the history of Spain and England in which geology is also involved. You will be able to find equally dramatic examples of contingency in the history of any place that interests you.

In 1588, Spain was the strongest power in Europe and controlled a great empire in both the Old World and New World with huge shipments of silver coming annually from Peru and Mexico. This enormous income was largely used to prosecute religious wars against the Protestants. Spain's powerful king, Philip II, who was particularly hostile to Protestant England, built a great fleet—the Spanish Armada—to cross the English Channel and bring the English back to Catholicism by force.

What contingencies went into Philip's plan? Even the fact that Philip was king of Spain at all was itself extremely unlikely, as we shall see later when considering how his father, Charles, came to be king before him. Without Philip's upbringing as a fervent Catholic, there would have been no motivation for an invasion of England. In addition, there would have been no need to build an Armada if there had not been an English Channel. In the sixteenth century, Spanish armies were the finest in Europe; no soldiers anywhere could stand up to a Spanish *tercio*. If England had been a peninsula rather than an island, Philip's army could have simply marched into England and taken it over. But there *is* an English Channel, and that natural moat has protected England from medieval times until the Second World War.

A geologist, of course, asks why the English Channel is there, and the answer based on recent research is fascinating. At some time during the Pleistocene Ice Age there was still a highland connecting England and France where the Straits of Dover now lie. Sea level was about 300 feet lower than today because of all the water locked up in ice on Canada and other great ice sheets. Glaciers covered much of Scandinavia and Britain. As they retreated, a large meltwater lake formed, trapped between the ice sheet to the north and the Dover highland, which was in fact an anticline—an upfold of the sedimentary rocks. At some point the lake level rose to a height where it spilled over the Dover dam, catastrophically eroding a great channel during what geologists call a megaflood. This megaflood was proposed four decades ago,[19] but only in 2007 was it fully confirmed,[20] when a detailed survey of the floor of the English Channel revealed the presence of streamlined former islands, carved by the megaflood waters, just like ones known from the glacial megaflood channels of central Washington. This is the hard science behind Steve Dutch's counterfactual-history speculation that we met in Chapter 6 suggesting that, had the last ice age been less icy, there might have been no English Channel, and Philip's Armada would not have been necessary.

The Armada was to sail from Spain to Flanders in 1588 to pick up a Spanish army, which it would transport to England for the invasion, but miscommunications and misunderstandings—common contingencies in warfare—interfered with the plan. Thus the Armada found itself anchored in the English Channel off Calais, awaiting the army. One night, with the wind blowing in a favorable direction, the English set fire to some old ships and sent them into the closely packed Spanish ships fleet. The Spanish captains, quickly setting sail, somehow all managed to avoid being struck and burned, but in the ensuing confusion the Armada was scattered and unable to pick up the army for the invasion of England. What with storms, English harassment, and general chaos, many of the ships and their crews were lost. The surviving ships tried to sail all the way around the British Isles and back to Spain. A few made it home, but the chance to invade England was lost. Perhaps we can conclude that a tiny geological contingency—the way the wind was blowing on a particular night—had something to do with why England remained Protestant, and why the United States was colonized by English speakers, not Spanish speakers. And yet it was only on that particular night that it mattered to history which way the wind was blowing; a chance event is evidently only significant if it occurs at what has been called a "critical juncture."

Contingency is everywhere in human history. A contingent, catastrophic, geological event produced the channel that made Philip's Armada necessary; the thinking of Martin Luther in Germany generated the Protestant Reformation that Philip was determined to reverse; the contingencies of Philip's life, thought, and beliefs led to the Armada being built; accidents of warfare kept the ships waiting to pick up his army; and contingent wind conditions allowed the English fireships to begin their destruction. But what was the chance that there would ever have been a Philip to set these events in motion? In our two final sections, let us look at the enormous contingency, the breathtakingly tiny odds, that Philip, or any of us human beings, would ever have come to be.

CONTINGENCY IN OUR LIVES

We can recognize continuities in our lives—like the aging trend from infancy to old age and cycles like day and night and the seasons. And yet we live our lives immersed in an ocean of contingency—accidents, the vagaries of illness or health, or a chance meeting that leads to conflict or friendship, or love. Even our existing at all is completely contingent and improbable to a degree that beggars belief. We can see this by looking at how Philip II came to exist and came to be king. And we can see it in the chance contingencies in each of our genealogies.

Philip II was king of Spain at the time of the Armada because of a remarkable set of contingencies a generation earlier. His father, the Emperor Charles V, was also king of Spain. Becoming king in 1516 and Holy Roman Emperor in 1519, Charles received a fabulous inheritance—an empire covering much of Europe and the newly discovered Americas. But a few years earlier no one could have predicted that he would ever inherit Spain or anything else. Charles's Spanish kingdom came through his mother, Juana, third child of Ferdinand and Isabel, the patron of Columbus. The crown should have gone to Prince Juan, their only son, but he died at age 19 and his only child was stillborn. Next in line was daughter Isabel, but she died giving birth to her only child, who in turn died in infancy. Juana became queen and her son Charles became king only because of four premature deaths from natural causes. Even in an age when the mortality risk during childbirth was high, the rise of Charles would have been surprising. It was these contingencies that brought Charles to the Spanish throne at the peak of its power and his son Philip in turn.

Philip's contingent kingship and the contingent fate of his Armada had enormous consequences for the unfolding of world history. Few people have significance at that level, but even the fact of our being here at all is highly contingent. Going back even one or two generations, the probability of any one of us ever being born would have been vanishingly small. Our mere existence meets the three requirements for contingency. Unless you have an identical twin, the birth of someone with exactly your genes was not only rare but unique as well; it was unpredictable; and it was significant at least at the level of family and friends. Each individual's rarity and improbability are quantifiable in at least a rough way. What we learn by estimating our improbability may give us insights into some of the deepest questions we can ask. This is a part of the human situation that I have never seen discussed, and it is a fitting way to close a book on the human situation.

My great-grandfather, Luis Fernández Álvarez, emigrated from Spain to Cuba at age 7 with his older brother, because their father had died in a fall from a balcony in Bilbao, and they had already lost their mother. Eventually he ended up as a doctor in Hawaii and California. I feel bad every time I think of the untimely death of my great-great-grandfather, but I also realize that without that contingent event, Luis would probably never have left Spain, there would never have been a California branch of our family, and I would not exist.

When I tell people about this, I sometimes hear the most remarkable stories of the contingencies that have allowed them to be here. Rudy Saltzer was a dear friend and a fine choral conductor; this is Rudy's story: His father was a soldier in the Russian army in World War I. Having been awake all night as a sentry, he volunteered for a second night's sentry duty to replace a sick friend and fell asleep—a capital offense. He was behind bars when a visiting officer passed by with the captain of the unit on a tour of inspection. The inspector noticed Rudy's father in the prison and asked about him. The captain said he had fallen asleep on sen-

Fig. 10-2. Luis Fernández Álvarez (1853–1937). My great-grandfather, who had to leave Spain as a child, after first his mother died and then his father. Had he not lost his parents at a young age, I would not exist.

try duty and was going to be shot in the morning. The inspector agreed that this was unfortunately necessary, and then he asked if the prisoner had been a good soldier. The captain said yes, he was a good soldier, with this one exception, of course. Then he added that there was something unusual about Rudy's father—he was Jewish. The inspector thought about this for a while, and then he said, "Isn't this night special for Jewish people? I think it's called Yom Kippur." The captain said yes, he thought he had heard of that. And then the inspector said, "Since he was a good soldier, and since this is a special day for Jewish people, let's forgive him, just this once." And a few years later, after Rudy's parents emigrated, Rudy was born in Los Angeles.

We all have contingencies that led to our being here, though not always as dramatic as Rudy's story. Think about the chances that your parents, your grandparents, and each couple in your ancestry, for a thousand million years back, would have met, would have had a child, and would have had that particular child who became one of your ancestors, or became you! Bill Bryson begins his magnificent telling of the Big History saga with exactly this point—that we are each almost unbelievably improbable.[21] Just how improbable are we?

How improbable are we?

Here are two ways to think about just how improbable each of us is: In the first method, think about your family tree—you, with two parents, four grandparents, eight great-grandparents and so on, another factor of two for each generation back into the past. Ten generations back you have about a thousand ancestors; 20 generations back a million, 30 generations a thousand million, and so on. Actually we should not say a million *ancestors* but rather a million boxes at that level in the tree, because you are descended from particular individuals via multiple paths. This is clear when you realize that back beyond the Renaissance there are more boxes—*far* more boxes—at that level in your family tree than there were people alive at the time. No wonder people interested in their genealogy seldom make a complete family tree back more than a few generations!

So here's the point. The sex of a child is established essentially at random during conception, depending on whether the successful sperm cell carries an X or a Y chromosome. Now if any single one of your ancestors in the gazillions of boxes back to the beginning of multicellular life, about a billion years ago, had been of the opposite sex, then that individual could not occupy that box, and you would not exist. This is a vivid demonstration of the contingency of our birth and gives a sense of our very existence hanging from the most tenuous web of chance going all the way back to the beginning of life. Some people see this as almost terrifying. Others see it as making them really special.

So let's turn to the second method to see if we can get a rough estimate of how probable—or improbable—we are. If we ask how many people will be born into the next generation worldwide, the answer is on the order of a billion, about 10^9. If we ask how many individuals *might* be born into that generation, considering the number of eggs and sperm involved, the answer is about 10^{25} in very rough numbers.[22]

What do those numbers mean? Does 10^{25} sound all that much larger than 10^9? Most people don't often stop to think about the actual meaning of numbers written exponentially. So here is a way to visualize it: If you take grains of fine sand, 10^9 is a double handful, but 10^{25} grains of fine sand would fill ten Grand Canyons! We who are alive today are the handful of people who were actually born, and the ten Grand Canyons full of sand grains represent all those possible people who never got to live.

And it only gets worse if we consider multiple generations, with about 10^{50} possible individuals over two generations, 10^{75} over three generations, 10^{100} over four generations, and so on. These at first may sound like astronomical numbers, but in fact they are not. . . . They are *hyper*-astronomical numbers! The number 10^{100} is far, far larger than the estimated number of elementary particles in the visible universe (about 10^{80}). Maybe seeing 10^{100} written out would make it clear how big a number it is:[23]

10,000,000,000,000,000,000,000,000,000,000,000,000,000,000,0
00,000,000,000,000,000,000,000,000,000,000,000,000,000,000,0
00,000,000,000.

This is the number of individual humans who had the potential to be born into our generation at the time when our great-great-grandparents were having *their* children. We are the few who *were* born. We can only conclude that each of us, and every individual we ever meet, has been a winner in the most ruthless game of chance ever concocted. We are all winners against hyper-astronomical odds!

Almost 14 billion years of Cosmic history, more than 4 billion years of Earth and Life history, a couple of million years of Human history, all of it constrained by the laws of Nature but playing out in an entirely unpredictable way because of countless contingencies—this history has produced the human situation in which we live. We few, we fortunate few, are the ones who have inherited this world and this situation, and it is our actions that will influence the next chapters in the unfolding journey of Big History.

APPENDIX I:

FURTHER RESOURCES

FOR THOSE INTERESTED in deepening their understanding of Big History and its component regimes, there are many places to turn. Here are some of the books and other resources that have influenced my thinking.

Big History in General

Bill Bryson's best-selling 2003 explanation of many fields of science is in fact an excellent overview of Big History by a fine writer: *A short history of nearly everything*, New York, Broadway Books, 544 p.

Another good place to start is with two concise, thoughtful books about the nature of Big History by Fred Spier: 1996, *The structure of Big History*, Amsterdam, Amsterdam University Press, 113 p.; 2010, *Big History and the future of humanity*, Chichester, John Wiley and Sons, 272 p.

David Christian, the founder of the field of Big History, has written both a long overview of what has happened in all of the past and a brief summary of all of Big History: 2004, *Maps of time. An introduction to Big History*, Berkeley, University of California Press, 642 p.; 2008, *This fleeting world: A short history of humanity*, Great Barrington, Mass., Berkshire, 112 p. Cynthia Brown has also written a book reviewing all of the past: 2007, *Big History: from the Big Bang to the present*, New York, New Press, 288 p.

David Christian, Cynthia Brown, and Craig Benjamin have recently published the first textbook of Big History: 2014, *Big History: Between nothing and everything*, New York, McGraw Hill, 332 p.

Carl Sagan's *Cosmos* was perhaps the first modern presentation of what was to become Big History. It is available both as television program segments on the Web and in book form: 1980, *Cosmos*, New York, Random House, 365 p.

Progress in understanding Big History is strongly tied to the ability to date events in all four of its regimes, and a good explanation of several of the most important dating techniques is given by Matthew Hedman, 2007, *The age of everything: How science explores the past*, Chicago, University of Chicago Press, 249 p.

Two extensive compilations of articles about Big History have been edited by a Russian-U.S. team: L. E. Grinin, A. V. Korotayev, and B. H. Rodrigue, 2011, *Evolution: A Big History perspective*, Volgograd (Russia), Uchitel Publishing House, 303 p.; B. H. Rodrigue, L. E. Grinin, and A. V. Korotayev, 2011, *From Big Bang to galactic civilizations: A Big History anthology*, v. 1: *Our place in the universe—An introduction to Big History*, Delhi, Primus Books, 357 p. (further volumes to appear).

Paleoanthropologists Kathy Schick and Nick Toth have built a website that takes you *From the Big Bang to the World Wide Web*: http://www.bigbangtowww.org/.

The International Big History Association holds conferences, publishes a monthly newsletter, and welcomes new members: http://www.bigbangtowww.org/.

The Big History Project, led by David Christian and Bill Gates, has made available a great deal of information about all aspects of the past: https://www.bighistoryproject.com/home.

Videos of several of my lectures about Big History, and those of David Christian and other Big Historians can be found on YouTube.

COSMIC REGIME

Wonderful pictures of the Cosmos, with a new one each day, are available on the Web at *Astronomy Picture of the Day*: http://apod.nasa .gov/apod/astropix.html.

My favorite book about the Big Bang is by Alan Guth, the discoverer of cosmic inflation: 1997, *The inflationary universe*, New York, Vintage, 358 p. Another good introduction to the Big Bang is by Steven Weinberg: 1993, *The first three minutes: A modern view of the origin of the universe*, New York, Basic Books, 203 p. The detailed structure of cosmic background radiation is explained by its discoverer, George Smoot, with Keay Davidson: 1993, *Wrinkles in time*, New York, Morrow, 331 p.

Lisa Randall gives a thorough discussion of the nature of dark matter and the difficult problem of detecting it, and she speculates about a possible connection to mass extinctions: 2015, *Dark matter and the dinosaurs*, New York, HarperCollins, 412 p. In fact, all of her books—*Warped Passages* (2005), *Knocking on Heaven's Door* (2011), *Higgs Discovery* (2012), and *Dark Matter and the Dinosaurs* (2015)— are an excellent path to a deeper understanding of the Cosmos and Cosmic history. In another interesting book at the cutting edge of astronomy, Anna Frebel explains the search for the oldest stars in our galaxy: 2015, *Searching for the oldest stars: Ancient relics from the early universe*, Princeton, Princeton University Press, 302 p.

A profound question at the heart of cosmology is why the universe works the way it does—how and when were the mathematical laws of Nature embedded in the Cosmos? Martin Rees tackles that question in a fascinating and influential book that supports the concept of the multiverse: 2003, *Just six numbers: The deep forces that shape the universe*, New York, Basic Books, 176 p. Paul Davies has given us a number of thoughtful books about the deepest questions of cosmology: 1983, *God and the new physics*, New York, Simon and Schuster, 255 p.; 1988, *The cosmic blueprint: New discoveries in nature's creative ability to order the universe*, New York, Simon and Schuster, 224 p.; 1992, *The mind of God: The scientific basis for a rational world*, New York, Simon and Schuster, 254 p.; and 2007, *The cosmic jackpot: Why our universe is just right for life*, Boston, Houghton Mifflin, 315 p.

Timothy Ferris has given us interesting views of astronomy: 1988, *Coming of age in the Milky Way*, New York, Morrow, 495 p.; 1992, *The mind's sky: Human intelligence in a cosmic context*, New York, Bantam, 281 p. The classic scholarly study of the Copernican Revolution is by Thomas Kuhn: 1957, *The Copernican Revolution*, Cambridge, Mass., Harvard University Press, 297 p. Jack Repcheck has written an excellent short biography of Copernicus: 2007, *Copernicus' secret: How the scientific revolution began*, New York, Simon and Schuster, 239 p. The equivalent treatment of Newton is by James Gleick: 2003, *Isaac Newton*, New York, Pantheon, 272 p.

An excellent introductory textbook in astronomy is by Jay Pasachoff and Alex Filippenko, 2007, *The Cosmos*, Belmont, Calif., Brooks/Cole, 500 p.

EARTH REGIME

An Earth Picture of the Day is available on the Web: http://epod .usra.edu/.

Perhaps the first notable geologist to write about Earth history in the broader context of what was to become Big History was Preston Cloud: 1978, *Cosmos, Earth, and man: A short history of the universe*, New Haven, Yale University Press, 372 p., and 1988, *Oasis in space: Earth history from the beginning*, New York, W. W. Norton, 508 p. A similar approach was taken by Cesare Emiliani: 1992, *Planet Earth: Cosmology, geology, and the evolution of life and environment*, Cambridge, Cambridge University Press, 717 p.

My two previous books place aspects of Earth history into a much broader Big History context: Walter Alvarez, 1997, *T. rex and the Crater of Doom*, Princeton, Princeton University Press, 185 p., and 2009, *The Mountains of Saint Francis*, New York, W. W. Norton, 304 p.

The excellent writer John McPhee has given us five books about geology and geologists, which are collected in a single volume: 1998, *Annals of the former world*, New York, Farrar, Straus and Giroux, 696 p. And from geologist Marcia Bjornerud comes a good personal introduction to the study of Earth: 2005, *Reading the rocks: The autobiography of the Earth*, Cambridge, Mass., Westview Press, 237 p.

Accessible, short biographies of three of the pioneers of geology—Nicolaus Steno, James Hutton, and William Smith—are helpful in understanding how scientists first came to realize that Earth has had a very long history and that it is recorded in rocks: Alan Cutler, 2003, *The seashell on the mountaintop: A story of science, sainthood, and the humble genius who discovered a new history of the Earth*, New York, Dutton, 228 p.; Jack Repcheck, 2003, *The man who found time*, Cambridge, Mass., Perseus, 247 p.; and Simon Winchester, 2001, *The map that changed the world: William Smith and the birth of modern geology*, New York, HarperCollins, 329 p.

The evolution of Earth and how it became the abode for life and for humanity is explained by Charles Langmuir and Wally Broecker: 2012, *How to build a habitable planet: The story of Earth from the Big Bang to humankind*, Princeton, Princeton University Press, 718 p. This is also the approach of Jonathan Lunine: 2013, *Earth: Evolution of a habitable world*, 2nd edition, Cambridge, Cambridge University Press, 327 p. For a more technical study of Earth history, see Kent Condie: 2005, *Earth as an evolving system*, Amsterdam, Elsevier, 447 p.

Brent Dalrymple has given a detailed explanation of how geologists go about dating events in Earth's past: 1991, *The age of the Earth*, Stanford, Calif., Stanford University Press, 474 p.

The geological history of two mountain ranges—the Andes and the Apennines—is explained by Simon Lamb, 2004, *Devil in the mountain: A search for the origin of the Andes*, Princeton, Princeton University Press, 335 p.; and Walter Alvarez, 2009, *The Mountains of Saint Francis*, New York, W. W. Norton, 304 p. A number of useful books about the geology of particular places, including the *Roadside Geology* series, covering several U.S. states, are available from the Mountain Press of Missoula, Montana: http://mountain-press.com/.

The history of climate is of much importance in this time of impending climate change, and interesting accounts have been given by Brian Fagan: 1999, *Floods, famines, and emperors: El Niño and the fate of civilization*, New York, Basic Books, 346 p., and 2008, *The great warming: Climate change and the rise and fall of civilizations*, New York, Bloomsbury Press, 282 p.; by Richard Muller and Gordon MacDonald, 2000, *Ice ages and astronomical causes*, New York, Springer, 318 p.; and by Lynn Ingram and Frances Malamud-Roam, 2013, *The West without water: What past floods, droughts, and other climatic clues tell us about tomorrow*, Oakland, University of California Press, 280 p.

My favorite textbook of geology, by Steve Marshak, comes in both longer and shorter versions: 2015, *Earth: Portrait of a planet*, 5th edition, New York, W. W. Norton, 984 p.; 2016, *Essentials of geology*, 5th edition, New York, W. W. Norton, 575 p.

LIFE REGIME

We are fortunate to have a number of excellent books about paleontology by leading scientists in the field. The very difficult search for the oldest fossils is described by Andy Knoll: 2003, *Life on a young planet: The first three billion years of evolution on Earth*, Princeton, Princeton University Press, 277 p. Richard Fortey has given us 2001, *Trilobite: Eyewitness to evolution*, New York, Vintage, 284 p. The discovery of the forms transitional between fish and land animals is recounted by Neil Shubin: 2008, *Your inner fish: A journey into the 3.5-billion-year history of the human body*, New York, Random House, 240 p. Among many books about dinosaurs is Michael Novacek's 2003, *Time traveler: In search of dinosaurs and other fossils from Montana to Mongolia*, New York, Farrar, Straus and Giroux, 380 p. The whole field of life history is discussed by Don Prothero: 2007, *Evolution: What the fossils say and why it matters*, New York, Columbia University Press, 408 p.; and by Nick Lane: 2009, *Life ascending: The ten great inventions of evolution*, New York, W. W. Norton, 344 p.

Many books have treated the question of mass extinctions. The extinction 66 million years ago that eliminated the dinosaurs was discussed early on by Dave Raup: 1981, *Extinction: Bad genes or bad luck?*, New York, W. W. Norton, 210 p.; by John Noble Wilford: 1985, *The riddle of the dinosaur*, New York, Knopf, 304 p.; by Kenneth Hsü: 1986, *The great dying: Cosmic catastrophe, dinosaurs, and the theory of evolution*, San Diego, Harcourt Brace Jovanovich, 292 p.; by Richard Muller: 1988, *Nemesis, the death star: The story of a scientific revolution*, New York, Weidenfeld and Nicolson, 193 p.; by Bill Glen: 1994, *The mass-extinction debates: How science works in a crisis*, Stanford, Calif., Stanford University Press, 376 p.; by James Powell: 1998, *Night comes to the Cretaceous*, New York, W. H. Freeman, 250 p.; in my book, Walter Alvarez: 1997, *T. rex and the Crater of Doom*, Princeton, Princeton University Press, 185 p.; and by Charles Frankel: 1999, *The end of the dinosaurs: Chicxulub Crater and mass extinctions* (translation from the French), Cambridge, Cambridge University Press, 223 p.

The even greater mass extinction at the end of the Permian, 252 million years ago, is treated by Michael Benton: 2003, *When life nearly died: The greatest mass extinction of all time*, London, Thames and Hudson, 336 p.; and by Peter Ward: 2004, *Gorgon: Paleontology, obsession, and the greatest catastrophe in Earth's history*, New York, Viking, 257 p. The mass extinction that is happening now, due largely to human activities, is the subject of Tony Barnosky's 2014, *Dodging extinction: Power, food, money, and the future of life on Earth*, Oakland, University of California Press, 240 p.; and Elizabeth Kolbert's 2014, *The sixth extinction: An unnatural history*, New York, Holt, 319 p.

Turning from paleontology to molecular genetics as a tool for investigating life history, the field is so new that general books are hard to find. Molecular genetics is very good at determining the evolutionary relationships between taxa, that is, for drawing an accurate tree of life, and this is the aim of Joel Cracraft and Michael Donoghue, eds.: 2004, *Assembling the tree of life*, Oxford, Oxford University Press, 576 p. To go further and attach dates to the branches of the tree requires assumptions about rates of genetic change; an attempt to do this is by Blair Hedges and Sudhir Kumar, eds.: 2009, *The timetree of life*, Oxford, Oxford University Press, 551 p. The link between molecular genetics and paleontology, investigating how the proteins coded by DNA give rise to the forms of animals and plants in life history, is the field of evolutionary developmental biology. An excellent introduction to this topic is by Sean Carroll: 2005, *Endless forms most beautiful: The new science of evo devo and the making of the animal kingdom*, New York, W. W. Norton, 550 p.

A good textbook of paleontology and life history is by Richard Cowen: 2013, *History of life*, 5th edition, Hoboken, NJ, Wiley, 302 p.

HUMANITY REGIME

Yuval Noah Harari's 2015, *Sapiens: A brief history of humankind*, New York, HarperCollins, 443 p., is a remarkable study of humanity, full of startling insights that may completely change the way you think about our species and its history.

The invention of writing divides the Humanity regime into two parts. Prewriting human history is largely the domain of archaeologists and anthropologists. For the work of paleoanthropologists in finding fossils of early humans, see Don Johanson's classic with Maitland Edey: 1981, *Lucy, the beginning of humankind*, New York, Simon and Schuster, 409 p. Two other interesting books on this topic are by Carl Swisher, Garniss Curtis, and Roger Lewin: 2000, *Java Man: How two geologists' dramatic discoveries changed our understanding of the evolutionary path to modern humans*, New York, Scribner, 256 p.; and by Roger Lewin: 1987, *Bones of contention: Controversies in the search for human origins*, New York, Simon and Schuster, 348 p.

Kathy Schick and Nick Toth, the leading researchers on stone tools, have written about that topic: 1993, *Making silent stones speak*, New York, Simon and Schuster, 352 p.

Moving to a more general view of prehistory, a particularly fine overview is given by Nicholas Wade: 2006, *Before the dawn*, London, Penguin, 314 p. Also useful is Colin Renfrew's 2007, *Prehistory: The making of the human mind*, New York, The Modern Library, 219 p. For an in-depth study of the domestication of fire, see Johan Goudsblom: 1992, *Fire and civilization*, London, Penguin, 247 p. A brief introduction to the spread of humanity over the globe based on molecular genetics is by Spencer Wells: 2007, *Deep ancestry*, Washington, D.C., National Geographic Society, 247 p.

Two useful textbooks are by Richard Klein: 2009, *The human career: Human biological and cultural origins*, Chicago, University of Chicago Press, 989 p.; and by Chris Scarre, ed.: 2009, *The human past: World prehistory and the development of human societies*, 2nd edition, London, Thames and Hudson, 784 p.

Our knowledge of human history changes character and becomes much more detailed with the appearance of writing. Many studies of written history are specialized and focus on particular groups and times. "World History" is a current effort to achieve a broader understanding of the human past, and an overview of the field is given by Patrick Manning: 2003, *Navigating world history: Historians create a global past*, New York, Palgrave Macmillan, 425 p.

There are many books on world history. Among those that seek patterns and deep understanding, rather than simply recounting what has happened, are books by John and William McNeill: 2003, *The human web: A bird's-eye view of world history*, New York, W. W. Norton, 350 p.; by William McNeill: 1963, *The rise of the West*, Chicago, University of Chicago Press, 829 p., and 1992, *The global condition: Conquerers, catastrophes, and community*, Princeton, Princeton University Press, 171 p.; by William Ruddiman: 2005, *Plows, plagues, and petroleum: How humans took control of climate*, Princeton, Princeton University Press, 202 p.; by Paul Colinvaux: 1980, *The fates of nations: A biological theory of history*, New York, Simon and Schuster, 383 p.; by Steven Pinker: 2011, *The better angels of our nature: Why violence has declined*, New York, Viking, 802 p.; and by Daron Acemoglu and James A. Robinson: 2012, *Why nations fail: The origins of power, prosperity and poverty*, New York, Crown Publishers, 529 p.

The character of history

Steven Jay Gould's study of the history of geology: 1987, *Time's arrow, time's cycle: Myth and metaphor in the discovery of geological time*, Cambridge, Mass., Harvard University Press, 222 p., led me to think about a different dichotomy—continuity versus contingency. There is a rich literature on complexity and chaos theory and its relevance to contingency, and these are good places to start: James Gleick: 1987, *Chaos: Making a new science*, New York, Viking, 352 p.; Mitchell Waldrop, 1992: *Complexity: The emerging science at the edge of order and chaos*, New York, Simon and Schuster, 380 p.; and John Briggs and David Peat: 1989, *Turbulent mirror: An illustrated guide to chaos theory and the science of wholeness*, New York, Harper and Row, 222 p.

For more detailed treatments by scientists who have pioneered the study of chaos, see Benoit Mandelbrot: 1983, *The fractal geometry of nature*, New York, W. H. Freeman, 468 p.; Ilya Prigogine and Isabelle Stengers: 1984, *Order out of chaos: Man's new dialogue with nature*, Toronto, Bantam, 349 p.; and Edward Lorenz: 1993, *The essence of chaos*, Seattle, University of Washington Press, 227 p. John Lewis Gaddis has written a thoughtful book about how historians work and about the character of the history they are attempting to understand: 2002, *The landscape of history: How historians map the past*, Oxford, Oxford University Press, 192 p., which has a particularly fine discussion of contingency in Chapter 6.

Counterfactual history is helpful in thinking about contingency. For two collections of essays in counterfactual history, focusing on military situations, see Robert Cowley, ed.: 1999, *What if?: The world's foremost military historians imagine what might have been*, New York, Berkley Books, 395 p., and 2001, *More what if?: Eminent historians imagine what might have been*, New York, Putnam, 427 p. Reaching further back into Big History, Neil Comins has written two books about counterfactual history and Earth: 1993, *What if the moon didn't exist?: Voyages to Earths that might have been*, New York, HarperCollins, 315 p., and 2010, *What if the earth had two moons?*, New York, St. Martin's Press, 288 p.

Finally, since Big History is immersed in a matrix of time, there are several interesting books about different aspects of this central and quite mysterious feature of the human situation: Alan Lightman: 1993, *Einstein's dreams*, New York, Warner Books, 179 p.; Richard Morris: 1985, *Time's arrows: Scientific attitudes toward time*, New York, Simon and Schuster, 240 p.; James Ogg, Gabi Ogg, and Felix Gradstein: 2008, *The concise geologic time scale*, Cambridge, Cambridge University Press, 177 p.; Sean Carroll: 2010, *From eternity to here: The quest for the ultimate theory of time*, New York, Penguin, 438 p.; Claudia Hammond: 2013, *Time warped: Unlocking the mysteries of time perception*, New York, Harper, 342 p.; and Richard A. Muller: 2016, *Now: the physics of time*, New York, W. W. Norton, 364 p.

APPENDIX 2:

FIGURE SOURCES

Chapter 1: *Big History, the Earth, and the Human Situation*

Fig. 1-1: Photo by the author.
Fig. 1-2: Image from NASA, http://www.nasa.gov/sites/default/files/
images/712129main_8247975848_88635d38a1_0.jpg.

Chapter 2: *From the Big Bang to Planet Earth*

Fig. 2-1: Top: Hubble Heritage Team (AURA/ STScI/ NASA),
http://apod.nasa.gov/apod/ap010520.html. Middle: NASA, ESA,
Hubble Heritage Team (STScI/AURA), and W. P. Blair (JHU)
et al., http://apod.nasa.gov/apod/ap140128.html. Bottom: NASA,
ESA, G. Illingworth (UCO/Lick & UCSC), R. Bouwens (UCO/
Lick & Leiden U.), and the HUDF09 Team, http://apod.nasa.gov/
apod/ap091209.html.
Fig. 2-2: Photograph by Margaret Harwood, courtesy AIP Emilio
Segre Visual Archives.
Fig. 2-3: SOHO-EIT Consortium, ESA, NASA, http://apod.nasa
.gov/apod/ap101018.html.
Fig. 2-4: NASA, ESA, J. Hester, A. Loll (ASU); Acknowledgment:
Davide De Martin (Skyfactory), http://apod.nasa.gov/apod/ap111225
.html.

Fig. 2-5: NASA, ESA, Hubble Heritage (STScI/AURA)/Hubble-Europe Collaboration. Acknowledgment: D. Padgett (GSFC), T. Megeath (University of Toledo), B. Reipurth (University of Hawaii), http://apod.nasa.gov/apod/ap151218.html.

Fig. 2-6: NASA, ESA, N. Smith (University of California, Berkeley) et al., and the Hubble Heritage Team (STScI/AURA), http://apod .nasa.gov/apod/ap070425.html. A wonderful zoomable, color image of the Carina Nebula by Nathan Smith at University of California, Berkeley, can be explored at http://hubblesite.org/newscenter/ archive/releases/2007/16/image/a/format/zoom/.

Chapter 3: *Gifts from the Earth*

Fig. 3-1: Photo by the author.

Fig. 3-2: Satellite image from U.S. Geological Survey Department of the Interior/USGS, annotated by the author.

Fig. 3-3: Photo by Prof. Lung S. Chan, University of Hong Kong.

Fig. 3-4: Photo by the U.S. Geological Survey, http://volcanoes.usgs .gov/vsc/images/image_mngr/500-599/img523.jpg.

Fig. 3-5: Photo by the author.

Chapter 4: *A Planet with Continents and Oceans*

Fig. 4-1: On a thesis-inspection trip with Jack Lockwood. Photo by the author.

Fig. 4-2: © 2016 Colorado Plateau Geosystems, Inc. Reconstruction by Prof. Ron Blakey at http://jan.ucc.nau.edu/rcb7/300moll.jpg, labeled by the author. Many of Ron Blakey's reconstructed maps are available at http://jan.ucc.nau.edu/~rcb7/.

Fig. 4-3: Drawing by the author.

Fig. 4-4: Photo by the author, previously published as Fig. 1 in Leitão, H., and Alvarez, W., 2011, The Portuguese and Spanish voyages of discovery and the early history of geology: Geological Society of America Bulletin, v. 123, no. 7-8, p. 1219–1233.

Chapter 5: *A Tale of Two Mountain Ranges*

Fig. 5-1: Photo by the author. For a more detailed description, see Alvarez, W., 2009, The Mountains of Saint Francis, New York, W. W. Norton, Ch. 9.

Chapter 6: *Remembering Ancient Rivers*

Fig. 6-1: Map drawn by the author.
Fig. 6-2: Photograph by A. Loeffler. Image from the Library of Congress, https://www.loc.gov/item/97516175/.
Fig. 6-3: Lithograph published in 1855 by Herrman J. Meyer.
Fig. 6-4: Photo by the author.
Fig. 6-5: The figure is from a masterpiece of nineteenth-century geology: Gilbert, G. K., 1890, Lake Bonneville: Monographs of the U.S. Geological Survey, v. 1, Washington, D.C., Government Printing Office, 438 p., Fig. 21 on p. 98.
Fig. 6-6: Hydraulic gold mining near Dutch Flats, California, C. P. R. R. Library of Congress: American Memory, History of the American West, 1860–1920: Photographs from the Collection of the Denver Public Library.

Chapter 7: *Your Personal Record of Life History*

Fig. 7-1: Drawing by the author.

Chapter 8: *The Great Journey*

Fig. 8-1: Map drawn by the author. Topographic base from Smith and Sandwell global digital topographic map, http://topex.ucsd .edu/marine_topo/jpg_images/topo5.jpg.
Fig. 8-2: Location 25° 45.8′N, 12° 10.5′E, photo by the author.
Fig. 8-3: Modified by the author, on a new projection, from Wells, 2007, op. cit., and the Genographic Project, https://genographic .nationalgeographic.com/human-journey/.

Chapter 9: *Being Human*

Fig. 9-1: Photo by the author.
Fig. 9-2: NASA, http://apod.nasa.gov/apod/ap110518.html.
Fig. 9-3: Photo by the author.

Chapter 10: *Contingencies*

Fig. 10-1: Painting created for NASA by Donald E. Davis, 1994, http://www.jpl.nasa.gov/releases/98/yucatan.html.
Fig. 10-2: Author's family photograph.

* * * * *

The four-panel evocation of the regimes of Big History that intro-duces each section of this book was created by Nancy Crowe, a graphic designer with a background in Big History.

Acknowledgments

M Y INTEREST IN Big History began with geological fieldwork in many parts of the world in the company of Milly Alvarez. Fred Spier first told me about the emerging field of Big History and put me in touch with its founder, David Christian, with whom I've had many wonderful discussions.

Many of the ideas in this book were developed in discussions with close colleagues. *At Berkeley:* Raphael Bousso, Jiggs Davis, Chris Engberg, Olga García Moreno, Dan Karner, Henrique Leitão, Rich Muller, Mark Richards, Roland Saekow, and David Shimabukuro. *At the Geological Observatory of Coldigioco, in Italy:* Philippe Claeys, Christian Koeberl, Paul Kopsick, Paula Metallo, Sandro Montanari, Birger Schmitz, and Jan Smit. *At the Stone Age Institute in Indiana:* Kathy Schick and Nick Toth. *At Microsoft Research in Redmond, Washington:* Donald Brinkman, Bill Crow, Daron Green, Tony Hey, Rane Johnson, Lori Ada Kilty, Javier Porras Luraschi, Kal Viswanathan, Bob Walter, Curtis Wang, and Mike Zyskowski. *At family gatherings:* Don Alvarez, Jean Alvarez, Helen Alvarez, and Andrew Harth.

A lot of stimulating ideas have come from discussions with the colleagues who founded the International Big History Association at Coldigioco in 2010: Craig Benjamin, Cynthia Brown, David Christian, Michael Dix, Daron Green, Lowell Gustafson, Sandro Montanari, Barry Rodrigue, Roland Saekow, David Shimabukuro, and Fred Spier.

Many good ideas have come from the students and auditors of the Big History course I gave for five years at Berkeley with graduate student instructors David Shimabukuro, Dylan Spaulding, Joanne Emerson, Ryan Kelly, and David Mangiante. Roland Saekow's remarkable computer-zoom time line of Big History, "ChronoZoom," also emerged from that class and received very generous funding and great technical expertise from Microsoft Research for its development.

The following colleagues have given valuable comments on one or more chapters: Albert Ammerman, Peter Bickel, Raphael Bousso, Carlos Camargo, Gabriel Gutierrez Alonso, Eldridge Moores, Damian Nance, Lisa Randall, Jack Repcheck, Birger Schmitz, Jeff Shreve, and Jann Vendetti.

Jack Repcheck shepherded this project from the beginning with much wise counsel, and when Jack was sadly no longer able to continue, Jeff Shreve did a fine job of filling his shoes. And I'm indebted to Nancy Crowe for designing the section headings with the four panels representing the regimes of Big History.

To all of these colleagues and friends, and many others I surely am inadvertently leaving out, many thanks for a great intellectual adventure!

Notes

Chapter 1: Big History, the Earth, and the Human Situation

1. The term *Tertiary* is now obsolete, and this level should be called the Cretaceous-Paleogene boundary, but I will use Tertiary here because that is what we called it at the time.
2. Alvarez, L. W., Alvarez, W., Asaro, F., and Michel, H. V., 1980, Extraterrestrial cause for the Cretaceous-Tertiary extinction: Experimental results and theoretical interpretation: Science, v. 208, p. 1095–1108; Smit, J., and Hertogen, J., 1980, An extraterrestrial event at the Cretaceous-Tertiary boundary: Nature, v. 285, p. 198–200.
3. Montanari, A., Hay, R. L., Alvarez, W., Asaro, F., Michel, H. V., Alvarez, L. W., and Smit, J., 1983, Spheroids at the Cretaceous-Tertiary boundary are altered impact droplets of basaltic composition: Geology, v. 11, p. 668–671.
4. Muir, J. M., 1936, Geology of the Tampico region, Mexico: Tulsa, Oklahoma, American Association of Petroleum Geologists, 280 p.
5. Hildebrand, A. R., Penfield, G. T., Kring, D. A., Pilkington, M., Camargo-Zanoguera, A., Jacobsen, S. B., and Boynton, W. V., 1991, Chicxulub Crater: A possible Cretaceous/Tertiary boundary impact crater on the Yucatán Peninsula, Mexico: Geology, v. 19, no. 9, p. 867–871.
6. Smit, J., Montanari, A., Swinburne, N. H. M., Alvarez, W., Hildebrand, A. R., Margolis, S. V., Claeys, P., Lowrie, W., and Asaro, F., 1992, Tektite-bearing, deep-water clastic unit at the Cretaceous-Tertiary boundary in northeastern Mexico: Geology, v. 20, no. 2, p. 99–103.
7. Alvarez, W., 1997, T. rex and the Crater of Doom, Princeton, Princeton University Press, 185 p.
8. Magritte's painting is called "Castle in the Pyrenees." The original is in the Israel Museum in Jerusalem, and many images are available on the Web.

Chapter 2: From the Big Bang to Planet Earth

1. Carl Sagan made this point in his unforgettable description of the Pale Blue Dot—the tiny speck of light that is Earth, in the 1990 *Voyager* image from 3.7 billion miles away. Both the image and Sagan's comment can be found by searching for "Pale Blue Dot" on the Web.

2. Carl Sagan also chose to center this story around Milton Humason rather than Edwin Hubble, in Cosmos, Part 10, "The Edge of Forever," which is available on YouTube.

3. This stellar fixism is eerily reminiscent of the long acceptance by geologists, at about the same time, of continental fixism—the mistaken idea that continents have never moved—which we will meet in the next chapter. Alan Guth, 1997, The inflationary universe, New York, Vintage, Ch. 3, gives a clear account of the static-universe episode in the history of cosmology.

4. Brush, S. G., 2001, Is the Earth too old? The impact of geochronology on cosmology, 1929–1952, *in* Lewis, C. L. E., and Knell, S. J., eds., The age of the Earth: From 4004 BC to AD 2002: Geological Society of London, Special Publication, v. 190, p. 157–175.

5. This was the value given by Arthur Holmes in 1927 and 1931 (Dalrymple, G. B., 1991, The age of the Earth, Stanford, Calif., Stanford University Press, p. 17). The current value is a little over 4,500 million years (Dalrymple, G. B., 2001, The age of the Earth in the twentieth century: A problem (mostly) solved, *in* Lewis, C. L. E., and Knell, S. J., eds., The age of the Earth: from 4004 BC to AD 2002: Geological Society of London, Special Publication, v. 190, p. 205–221).

6. Kirshner, R. P., 2004, Hubble's diagram and cosmic expansion: Proceedings of the National Academy of Sciences, v. 101, no. 1, p. 8–13.

7. The astronomer Robert Kirshner has explained the full importance of the 1929 paper by Hubble, and we could take it as applying to the whole Hubble-Humason joint effort: Kirshner, R. P., 2004, Hubble's diagram and cosmic expansion: Proceedings of the National Academy of Sciences, v. 101, no. 1, p. 8–13.

8. Rees, M. J., 2003, Just six numbers: The deep forces that shape the universe, New York, Basic Books, 176 p.

9. The last event in this opening chapter of Big History came at about 380,000 years after the initiation of the Big Bang. By then the temperature of the universe had cooled to the point where protons and electrons, previously all mixed together in what physicists call a plasma, were able to combine into electrically neutral atoms of hydrogen. Before that time, photons of light were scattered constantly by the electrically positive protons and negative electrons. After their combination into neutral atoms, light could travel indefinitely, so this is called the "surface of last scattering." Light emitted at that point has been stretched out into radio waves by the Hubble-Humason expansion of the universe and is the source of the cosmic background radia-

tion that provides one of the three main lines of evidence for the reality of the Big Bang.

10. Helium made up about 25% of the normal matter of the universe at the end of the Big Bang when measured by mass; when measured by number of atoms, helium constituted about 10% of the normal matter. This does not include the dark matter, about which much less is known (Randall, L., 2015, Dark matter and the dinosaurs, New York, HarperCollins, 412 p.).

11. Until a few years ago it seemed that the interplay between thermal expansion and gravity was all that was needed to explain the overall structure of the universe. But the recent discovery of "dark energy," in which expansion is speeding up, not slowing down, has brought a wonderful new mystery to cosmology.

12. Lindberg, D. C., 1992, The beginnings of western science, Chicago, University of Chicago Press, p. 287–290.

13. The Sun's surface temperature is 5778 kelvins; at its center the temperature reaches more than 15 million kelvins.

14. Tolstikhin, I. N., and Kramers, D., 2008, The evolution of matter, Cambridge, UK, Cambridge University Press, 521 p.; Hazen, R. M., Papineau, D., Bleeker, W., Downs, R. T., Ferry, J. M., McCoy, T. J., Sverjensky, D. A., and Yang, H., 2008, Mineral evolution: American Mineralogist, v. 93, no. 11–12, p. 1693–1720.

15. Walter, M. J., and Trønnes, R. G., 2004, Early Earth differentiation: Earth and Planetary Science Letters, v. 225, p. 253–269.

16. The word "nebula" means cloud and was applied long ago to any object in the night sky that appeared fuzzy. We now know that the Carina Nebula is very much larger than the Crab Nebula, and they formed by completely different processes.

17. Canup, R. M., 2004, Dynamics of lunar formation: Annual Review of Astronomy and Astrophysics, v. 42, p. 441–475.

18. Comins, N. F., 2010, What if the Earth had two moons?, New York, St. Martin's Press, 288 p.

Chapter 3: Gifts from the Earth

1. Three terms used in this chapter may be confusing: "Silicon" means element 14, with the symbol, Si. "Silica" is a crystal with two oxygen atoms for every silicon atom—SiO_2, which is the formula of the mineral quartz. "Silicate" refers to a mineral containing silicon and oxygen, such as olivine (Mg_2SiO_4), with a structure based on the "silicate tetrahedron," a single small silicon atom surrounded by four large oxygen atoms in a tetrahedral arrangement. ("Silicone" is an artificial material made of silicon, oxygen, carbon, and hydrogen; it is not an Earth material and is not relevant to our story.)

2. *Minerals*, each with its characteristic formula, occur as grains that go together to make up *rocks*. A good analogy is that a rock is like a fruitcake, and minerals are like the fruit and nuts and cake within it. Since most minerals are quite resistant to change, so are the rocks that are made of minerals.

3. At first it seems surprising that silicon is the basis of rocks while carbon is the basis of life. Carbon lies just above silicon in the periodic table. Both make four chemical bonds with adjoining atoms, and this similarity has led science fiction writers to envision life based on silicon, but that is not what happens. (An example of silicon-based-life science fiction is the *Star Trek* episode, "The Devil in the Dark," available on the Web. Don't believe it! Silicon is the basis of rocks, not of alternative life.) The reason is that carbon is a versatile element, bonding to hydrogen, oxygen, nitrogen, sulfur, and other carbon atoms, with both single and double bonds, and easily changes the atoms to which it bonds. This versatility makes it the perfect basis for the complex structures and dynamic processes of life. In contrast, silicon greatly prefers bonding to oxygen, making strong single bonds but no double bonds. This is needed to make long-lasting minerals like the quartz (SiO_2) that is abundant in continental crust and the olivine (Mg_2SiO_4) that dominates in the deep-Earth mantle.

4. Igneous rocks remember the temperature, pressure, and chemistry of the molten magma from which they crystallized. Sedimentary rocks remember the processes that deposited them—rivers, wind, or glaciers. Metamorphic rocks remember the stresses that deformed them and the temperature and pressure under which they recrystallized into new minerals. Geologists have invented many ways of extracting this historical information from rocks and compiling it into an ever more sophisticated understanding of our planet's past. Thus geologists have a motto: *"Ex libro lapidum historia mundi"*; that is, "Out of the book of rocks comes the history of the Earth."

5. This question is answered in detail in an excellent little book: Broecker, W. S., 1985, How to build a habitable planet, Palisades, N.Y., Eldigio Press, 291 p.; in its expanded revision: Langmuir, C. H., and Broecker, W., 2012, How to build a habitable planet: The story of Earth from the Big Bang to humankind, Princeton, Princeton University Press, 718 p.; and in Chapter 11 of Gill, R., 2015, Chemical fundamentals of geology and environmental science, 3rd ed., Chichester, UK, Wiley, 267 p.

6. Those rare elements include lithium, beryllium, boron, most of the elements with odd numbers of protons, and almost all the heavy elements.

7. Silicon is what makes possible these large mineral grains because it has four places where it can bond to another atom. Each of those bonds prefers to link to an oxygen, which can make two bonds. So giant networks grow, each silicon surrounded by four oxygens, and each oxygen linking two silicons, with smaller amounts of other elements joining in—mostly magnesium and iron. These networks are called silicate minerals, and they are the basis of pretty much all of Earth, except for the iron core. No matter how big a silicate mineral grain has grown, there are always silicon and oxygen atoms at

the edge, waiting to grab onto more oxygens and silicons. Silicate grains can thus become arbitrarily large, unless they are in an aggregate of grains—a rock—and run into another growing grain.

8. Weir, A., 2011, The Martian, New York, Broadway Books, 369 p., adapted as a motion picture, 2015.

9. Harmand, S., Lewis, J. E., Feibel, C. S., Lepre, C. J., Prat, S., Lenoble, A., Boes, X., Quinn, R. L., Brenet, M., Arroyo, A., Taylor, N., Clement, S., Daver, G., Brugal, J.-P., Leakey, L., Mortlock, R. A., Wright, J. D., Lokorodi, S., Kirwa, C., Kent, D. V., and Roche, H., 2015, 3.3-million-year-old stone tools from Lomekwi 3, West Turkana, Kenya: Nature, v. 521, no. 7552, p. 310–315.

10. Toth, N., and Schick, K., 2010, Hominin brain reorganization, technological change, and cognitive complexity, *in* Broadfield, D., Yuan, M., Schick, K., and Toth, N., eds., The human brain evolving: Papers in honor of Ralph L. Holloway: Gosport, Ind., Stone Age Institute Press, p. 293–312.

11. Nick and Kathy's remarkable Big History museum exhibit, *From the Big Bang to the World Wide Web*, is available online at http://www.bigbang towww.org/.

12. Hess, H. H., 1960, The evolution of ocean basins (preprint): Princeton University, Department of Geology; Hess, H. H., 1962, History of ocean basins, *in* Engel, A. E. J., James, H. L., and Leonard, B. F., eds., Petrologic studies: A volume in honor of A. F. Buddington: Geological Society of America, p. 599–620.

13. Burke, K., McGregor, D. S., and Cameron, N. R., 2003, African petroleum systems: Four tectonic aces in the past 600 million years: Geological Society of London Special Publication, v. 207, p. 21–60, estimate (their p. 25) the volume of the North-African-Arabian Cambro-Ordovician sandstones as 15±5 million km^3, while the area of the United States, including Alaska, is about 10 million km^2.

14. A huge quarry in Illinois that extracts sand for use in glassmaking, from Ordovician deposits about 455–460 million years old, can be seen on Bing Maps or Google Earth at 41° 20.529′N, 88° 52.636′W.

15. Acemoglu, D., and Robinson, J. A., 2012, Why nations fail, New York, Crown Publishers, 529 p.

Chapter 4: A Planet with Continents and Oceans

1. Diamond, J., 1998, Guns, germs and steel: The fates of human societies, New York, W. W. Norton, 480 p., Ch. 10.

2. Wright, J. K., 1925/1965, The geographical lore of the time of the Crusades: A study of medieval science and tradition in Western Europe, New York, Dover, 563 p.

3. Ptolemy, C., second century, Geography (translated and edited by Edward Luther Stevenson), New York, New York Public Library, 1932, 167 p.

4. Page, M., 2002, The first global village, Alfragide (Portugal), Casa das Letras, 277 p.; Rodrigues, J. N., and Devezas, T., 2007, Pioneers of globalization: Why the Portuguese surprised the world, Vila Nova de Famalicão (Portugal), Centro Atlântico, 271 p.

5. Leitão, H., and Alvarez, W., 2011, The Portuguese and Spanish voyages of discovery and the early history of geology: Geological Society of America Bulletin, v. 123, no. 7–8, p. 1219–1233.

6. Russell, P. E., 2000, Prince Henry "the Navigator": A life, New Haven, Yale University Press, 448 p.

7. There is also a historical novel about the times of Prince Henry: Slaughter, F. G., 1957, The mapmaker: New York, Doubleday, 320 p. Finally, and best of all, is Portugal's national epic poem, *The Lusíads*, by Luis Vaz de Camões, which tells in poetic language the story of the Portuguese explorations (*The Lusíads*, written about 1556, published 1572, transl. by White, L., 1997, Oxford, Oxford University Press, 258 p.). The experience of the fearsome, unknown ocean is conveyed in the story of the monster Adamastor, in Canto V.

8. Eco, U., 1998, Serendipities: Language and lunacy: San Diego, Harcourt Brace, p. 4–7. This interesting information comes from Gabriel Gutiérrez Alonso, a present-day geology professor at Salamanca.

9. Braudel, F., 1966, 1973, The Mediterranean and the Mediterranean world in the age of Phillip II (2 volumes), transl. by Reynolds, S., New York, Harper and Row, 1375 p.

10. There is now a rich literature on this topic, mostly since about 2000, for example: Murphy, J. B., Gutiérrez-Alonso, G., Nance, R. D., Fernández-Saurez, J., Keppie, J. D., Quesada, C., Strachan, R. A., and Dostal, J., 2006, Origin of the Rheic Ocean: Rifting along a Neoproterozoic suture?: Geology, v. 34, no. 5, p. 325–328.

11. Cutler, A., 2003, The seashell on the mountaintop: A story of science, sainthood, and the humble genius who discovered a new history of the Earth, New York, Dutton, 228 p.; Alvarez, W., 2009, The Mountains of Saint Francis, New York, W. W. Norton, Ch. 5.

12. Romm, J., 1994, A new forerunner for continental drift: Nature, v. 367, p. 407–408.

13. Wegener, A., 1912, Die Entstehung der Kontinente: Geologische Rundschau, v. 3, p. 276–292; Wegener, A., 1929, The origin of continents and oceans (translation of Die Entstehun der Kontinente und Ozeane, reprinted 1966), New York, Dover, 246 p.

14. van Waterschoot van der Gracht, W. A. J. M., et al., 1928, Theory of continental drift, Tulsa, American Association of Petroleum Geologists, 240 p.

15. Hess, H. H., 1960, The evolution of ocean basins (preprint): Princeton University, Department of Geology.

16. Gould, S. J., 1987, Time's arrow, time's cycle, Cambridge, Mass., Harvard University Press, 222 p.

17. Repcheck, J., 2003, The man who found time, Cambridge, Mass., Perseus, 247 p.
18. Wilson, J. T., 1966, Did the Atlantic close and then re-open?: Nature, v. 211, p. 676–681.
19. Nance, R. D., Murphy, J. B., and Santosh, M., 2013, The supercontinent cycle: A retrospective essay: Gondwana Research, v. 25, no. 1, p. 4–29.
20. Moores, E. M., 1991, Southwest U.S.–East Antarctica (SWEAT) connection: A hypothesis: Geology, v. 19, no. 5, p. 425–428; Dalziel, I. W. D., 1991, Pacific margins of Laurentia and East Antarctica—Australia as a conjugate rift pair: Evidence and implications for an Eocambrian supercontinent: Geology, v. 19, no. 6, p. 598–601; Hoffman, P. F., 1991, Did the breakout of Laurentia turn Gondwanaland inside-out?: Science, v. 252, p. 1409–1412.
21. Gutiérrez-Alonso, G., Fernandez-Suarez, J., Weil, A. B., Murphy, J. B., Nance, R. D., Corfu, F., and Johnston, S. T., 2008, Self-subduction of the Pangaean global plate: Nature Geoscience, v. 1, no. 8, p. 549–553.
22. Leitão, H., and Alvarez, W., 2011, The Portuguese and Spanish voyages of discovery and the early history of geology: Geological Society of America Bulletin, v. 123, no. 7–8, p. 1219–1233.
23. Alvarez, W., 1972, Rotation of the Corsica-Sardinia microplate: Nature Physical Science, v. 235, p. 103–105; Alvarez, W., Cocozza, T., and Wezel, F. C., 1974, Fragmentation of the Alpine orogenic belt by microplate dispersal: Nature, v. 248, p. 309–314; Alvarez, W., 1976, A former continuation of the Alps: Geological Society of America Bulletin, v. 87, p. 891–896.
24. Rosenbaum, G., Lister, G. S., and Duboz, C., 2002, Reconstruction of the tectonic evolution of the western Mediterranean since the Oligocene, Oligocene: Journal of the Virtual Explorer, v. 8, p. 107–130; Hinsbergen, D. J. J., Vissers, R. L. M., and Spakman, W., 2014, Origin and consequences of western Mediterranean subduction, rollback, and slab segmentation: Tectonics, v. 33, no. 4, p. 393–419. For an explanation of how this works, see the discussion of the rather similar toe of the Italian boot, Calabria, in Alvarez, W., 2009, The Mountains of Saint Francis, New York, W. W. Norton, 304 p., Ch. 14.
25. This map shows modern boundaries of Portugal and Spain, except for the Muslim Emirate of Granada, shown as it was in 1462, when Christian Spain under Ferdinand and Isabella began its conquest. At the time, Madrid was still a small town and Toledo was the capital. The westbound Alborán microcontinent has pushed up the mountain range of the Sierra Nevada but has stretched, thinned, and therefore subsided below sea level; the seafloor around Alborán Island is thus continental, not oceanic (Platt, J. P., Behr, W. M., Johanesen, K., and Williams, J. R., 2013, The Betic-Rif Arc and its orogenic hinterland; a review: Annual Review of Earth and Planetary Sciences, v. 41, p. 313–357.). The moving front of the microcontinent is after Rosenbaum, G., Lister, G. S., and Duboz, C., 2002, Reconstruction of the

tectonic evolution of the western Mediterranean since the Oligocene, Oligocene: Journal of the Virtual Explorer, v. 8, p. 107–130. The seafloor in the probable area of the Lisbon Earthquake of 1755 is shown by Duarte, J. C., Rosas, F. M., Terrinha, P., Schellart, W. P., Boutelier, D., Gutscher, M.-A., and Ribeiro, A., 2013, Are subduction zones invading the Atlantic? Evidence from the southwest Iberia margin: Geology, v. 41, no. 8, p. 839–842.

26. Duarte, J. C., Rosas, F. M., Terrinha, P., Schellart, W. P., Boutelier, D., Gutscher, M.-A., and Ribeiro, A., 2013, Are subduction zones invading the Atlantic? Evidence from the southwest Iberia margin: Geology, v. 41, no. 8, p. 839–842.

27. The Caribbean and Scotia island arcs may be similar situations with subduction infecting the Atlantic Ocean from the west, and the same process has been proposed for the ancient Rheic Ocean (Waldron, J. W. F., Schofield, D. I., Murphy, J. B., and Thomas, C. W., 2014, How was the Iapetus Ocean infected with subduction?: Geology, v. 42, no. 12, p. 1095–1098).

28. Leitão, H., and Alvarez, W., 2011, The Portuguese and Spanish voyages of discovery and the early history of geology: Geological Society of America Bulletin, v. 123, no. 7–8, p. 1219–1233.

Chapter 5: A Tale of Two Mountain Ranges

1. Alvarez, W., 2009, The Mountains of Saint Francis, New York, W. W. Norton, 304 p.

2. An interesting exception is the Stanford Alpine Archaeology Project of Patrick Hunt, researching the route by which Hannibal's army entered Italy during the Second Punic War: Hunt, P., 2007, Alpine archaeology, New York, Ariel, 157 p.

3. Alvarez, W., 2009, op. cit., Ch. 9.

4. Pääbo, S., 2014, Neanderthal Man: In search of lost genomes, New York, Basic Books, 275 p.

5. Fowler, B., 2000, Iceman: Uncovering the life and times of a prehistoric man found in an Alpine glacier, New York, Random House, 313 p.

6. Bernstein, P. L., 2005, Wedding of the waters, New York, W. W. Norton, p. 22.

7. Nicolson, M. H., 1959, Mountain gloom and mountain glory: The development of the aesthetics of the infinite: Ithaca, N.Y., Cornell University Press, p. 2.

8. The history of the chronological interpretation of the Old Testament is reviewed in Ch. 2 of Repcheck, J., 2003, The man who found time: Cambridge, Mass., Perseus, 247 p.

9. Nicolson, M. H., 1959, op. cit.

10. Cutler, A., 2003, The seashell on the mountaintop: New York, Dutton, 228 p.; Alvarez, W., 2009, The Mountains of Saint Francis: New York, W. W. Norton, 304 p., Ch. 5.

11. Winchester, S., 2001, The map that changed the world: New York, Harper-Collins, 329 p.
12. Dalrymple, G. B., 1991, The age of the Earth: Stanford, Calif., Stanford University Press, 474 p.; Hedman, M., 2007, The age of everything: How science explores the past: Chicago, University of Chicago Press, 249 p.
13. Alvarez, W., 2009, The Mountains of Saint Francis: New York, W. W. Norton, photograph at the top of p. 156, by Prof. Stefan Schmid.
14. Bailey, E. B., 1935 (reprinted 1968), Tectonic essays, mainly Alpine: Oxford, Oxford University Press, 200 p.
15. This chapter focuses on collisional mountain ranges like the Alps and the Appalachians, formed where two continents are colliding. There is a second kind of mountain range, also tied into the supercontinent cycle, which forms above a subduction zone on the edge of a continent or as an arc of volcanic islands. The classical example is the Andes: Lamb, S., 2004, Devil in the mountain: A search for the origin of the Andes: Princeton, Princeton University Press, 335 p.
16. Willett, S. D., Schlunegger, F., and Picotti, V., 2006, Messinian climate change and erosional destruction of the central European Alps: Geology, v. 34, p. 613–616.
17. As you may be getting used to by now, the story is more complicated in detail: Fischer, K. M., 2002, Waning buoyancy in the crustal roots of old mountains: Nature, v. 417, p. 933–936.

Chapter 6: Remembering Ancient Rivers

1. Hames, W. E., McHone, Gregory, J., Renne, P. R., and Ruppel, C., 2003, The Central Atlantic Magmatic Province; insights from fragments of Pangaea, Geophysical Monograph, Volume 136, American Geophysical Union, Washington, D.C., 267 p.
2. Bernstein, P. L., 2005, Wedding of the waters, New York, W. W. Norton, 448 p.
3. In fact, because of the westbound train schedule, much of the Erie Canal route may be traversed in darkness. The eastbound Lakeshore Limited offers a better way to see the canal country during the day.
4. Dutch, S. I., 2006, What if? The ice ages had been a little less icy?: Geological Society of America Abstracts with Programs, v. 38, no. 7, p. 73.
5. Comins, N. F., 2010, What if the Earth had two moons?, New York, St. Martin's Press, 288 p.
6. Cowley, R., 1999, What if?: The world's foremost military historians imagine what might have been, New York, Berkley Books, 395 p.; Cowley, R., 2001, What if?: Eminent historians imagine what might have been, New York, Putnam, 427 p.
7. What if? The ice ages had been a little less icy? By Steven I. Dutch (Geological Society of America Abstracts with Programs, v. 38, no. 7, p. 73):

"The What If? historical book series explores the impact of historical events by presenting counterfactual scenarios, or alternative histories. Most geological events are too remote in time and too indirect in effect for a counterfactual approach to be any more than science fiction. One exception is the Pleistocene, which was recent enough and had dramatic impacts on human history. In this counterfactual scenario, I assume that the North American ice sheets never extended far below the Canadian border, and the Scottish and Scandinavian ice sheets never merged. The impacts of this alternative history include:

"First, the Missouri River would not have been diverted into its present course and would probably have re-established its former drainage to Hudson's Bay. The United States would have acquired a much smaller Louisiana Purchase, with no water route for Lewis and Clark to follow to the Pacific Northwest. The western boundary with Canada might well be far south of its present latitude.

"Second, the Ohio, and probably the Teays paleo-river, would not have been established. The St. Lawrence watershed might have extended far down the west side of the Appalachians. The Thirteen Colonies might well have been hemmed in to the west and confined permanently to the Atlantic Coast. There would be no Great Lakes and no Erie Canal. Without the Ohio and Missouri Rivers furnishing easy east-west water transport, American history would have been profoundly different.

"Third, if the Scottish and Scandinavian ice sheets had not merged, the ancestral Rhine-Thames system would have flowed unhindered across the North Sea shelf, rather than seeking a new outlet to the west. There would be no English Channel, no defeat of the Spanish Armada, no water barrier to thwart Napoleon or Hitler. Britain might still be a formidable naval power, but with a land frontier as well, its cultural and political independence would have been far less secure. In conclusion, two relatively small changes in Pleistocene paleogeography would have profoundly changed Western history."

8. Whitmeyer, S. J., and Karlstrom, K. E., 2007, Tectonic model for the Proterozoic growth of North America: Geosphere, v. 3, no. 4, p. 220–259.

9. Dickinson, W. R., and Gehrels, G. E., 2009, U-Pb ages of detrital zircons in Jurassic eolian and associated sandstones of the Colorado Plateau: Evidence for transcontinental dispersal and intraregional recycling of sediment: Geological Society of America Bulletin, v. 121, p. 408–433.

10. Van Wagoner, J. C., Mitchum, R. M., Champion, K. M., and Rahmanian, V. D., 1990, Siliciclastic sequence stratigraphy in well logs, cores, and outcrops, American Association of Petroleum Geologists, Methods in Exploration Series, v. 7, 55 p.

11. Janecke, S. U., and Oaks, R. Q., Jr., 2011, New insights into the outlet conditions of late Pleistocene Lake Bonneville, southeastern Idaho, USA: Geosphere, v. 7, no. 6, p. 1369–1391.

12. Yeend, W. E., 1974, Gold-bearing gravel of the ancestral Yuba River, Sierra Nevada, California: U.S. Geological Survey Professional Paper, n. 772, p. 1–85; Cassel, E. J., Grove, M., and Graham, S. A., 2012, Eocene drainage evolution and erosion of the Sierra Nevada Batholith across Northern California and Nevada: American Journal of Science, v. 312, no. 2, p. 117–144.

13. Visible on Bing Maps or Google Earth at 39° 22′N, 120° 55′W.

Chapter 7: Your Personal Record of Life History

1. Pääbo, S., 2014, Neanderthal Man: In search of lost genomes, New York, Basic Books, 275 p.

2. Currently the most favored explanation of the Late Heavy Bombardment is called the Nice model, after the French Mediterranean town where it was developed: Gomes, R., Levison, H. F., Tsiganis, K., and Morbidelli, A., 2005, Origin of the cataclysmic Late Heavy Bombardment period of the terrestrial planets: Nature, v. 435, p. 466–469.

3. Ballard, R. D., 2000, The eternal darkness: A personal history of deep-sea exploration, Princeton, Princeton University Press, Ch. 6.

4. Martin, W., and Russell, M. J., 2003, On the origins of cells: A hypothesis for the evolutionary transitions from abiotic geochemistry to chemoautotrophic prokaryotes, and from prokaryotes to nucleated cells: Phil. Trans. R. Soc. Lond., B, v. 358, p. 59–85; Russell, M. J., Nitschke, W., and Branscomb, E., 2013, The inevitable journey to being: Philosophical Transactions Royal Society B, v. 368: 20120254.

5. Kelley, D. S., Karson, J. A., Blackman, D. K., Früh-Green, G. L., Butterfield, D. A., Lilley, M. D., Olson, E. J., Schrenk, M. O., Roe, K. K., Lebon, G. T., and Rivizzigno, P., 2001, An off-axis hydrothermal vent field near the Mid-Atlantic Ridge at 30°N: Nature, v. 412, p. 145–149; Früh-Green, G. L., Kelley, D. S., Bernasconi, S. M., Karson, J. A., Ludwig, K. A., Butterfield, D. A., Boschi, C., and Proskurowski, G., 2003, 30,000 years of hydrothermal activity at the Lost City Vent Field: Science (Washington), v. 301, no. 5632, p. 495–498.

6. Marshall, C., 2015, The origin of life: Lecture at the University of California Museum of Paleontology, March 31, 2015.

7. Sleep, N. H., Zahnle, K. J., Kasting, J. F., and Morowitz, H. J., 1989, Annihilation of ecosystems by large asteroid impacts on the early Earth: Nature, v. 342, no. 6246 (9 November), p. 139–142.

8. Fischer, A. G., 1984, Biological innovations and the sedimentary record, *in* Holland, H. D., and Trendall, A. F., eds., Patterns of change in Earth evolution: Berlin, Springer-Verlag, p. 145–157.

9. Donald Canfield has written what is in effect a little Big History of oxygen: Canfield, D. E., 2014, Oxygen: A four billion year history, Princeton, Princeton University Press, 196 p.

10. This was the great discovery of Lynn Margulis: Sagan (later Margulis), L., 1967, On the origin of mitosing cells: Journal of Theoretical Biology, v. 14, no. 3, p. 225–274.

11. King, N., 2004, The unicellular ancestry of animal development: Developmental Cell, v. 7, p. 313–325.

12. Knoll, A. H., 2004, Life on a young planet: The first three billion years of evolution on Earth, Princeton, Princeton University Press, Ch. 9.

13. Ward, P. D., and Brownlee, D., 2000, Rare Earth: Why complex life is uncommon in the universe, New York, Copernicus (Springer Verlag), 333 p.

14. The full cause of the Cambrian Explosion is still mysterious and controversial: Sperling, E. A., Frieder, C. A., Raman, A. V., Girguis, P. R., Levin, L. A., and Knoll, A. H., 2013, Oxygen, ecology, and the Cambrian radiation of animals: Proceedings of the National Academy of Sciences, v. 110, no. 33, p. 13,446–13,451.

15. Knoll, 2004, op. cit.

16. Shubin, N., 2008, Your inner fish: A journey into the 3.5-billion-year history of the human body, New York, Random House, 240 p.; Clack, J. A., 2009, The fin to limb transition: New data, interpretations, and hypotheses from paleontology and developmental biology: Annual Review of Earth and Planetary Sciences, v. 37, p. 163–179.

17. Winchell, A., 1886, Walks and talks in the geological field, New York, Chautauqua Press, p. 252. (I previously used this quote in T. rex and the Crater of Doom, 1997, Princeton, Princeton University Press, p. 57.)

18. Clemens, W. A., 1970, Mesozoic mammalian evolution: Annual Review of Ecology and Systematics, v. 1, no. 1, p. 357–390.

19. Renne, P. R., Deino, A. L., Hilgen, F. J., Kuiper, K. F., Mark, D. F., Mitchell, W. S., III, Morgan, L. E., Mundil, R., and Smit, J., 2013, Time scales of critical events around the Cretaceous-Paleogene boundary: Science, v. 339, no. 6120, p. 684–687. These authors used high-precision radiometric dating to demonstrate that the Chicxulub impact and the mass extinction were synchronous with an uncertainty of only 0.032 million years, confirming what Jan, Sandro, and I concluded from the Arroyo el Mimbral outcrop in northeastern Mexico (Ch. 1).

20. Schulte, P., Alegret, L., Arenillas, I., Arz, J. A., Barton, P., Bown, P. R., Bralower, T., Christeson, G. L., Claeys, P., Cockell, C. S., Collins, G. S., Deutsch, A., Goldin, T., Goto, K., Grajales-Nishimura, J. M., Grieve, R., Gulick, S., Johnson, K. D., Kiessling, W., Koeberl, C., Kring, D. A., MacLeod, K. G., Matsui, T., Melosh, J., Montanari, A., Morgan, J. V., Neal, C. R., Nichols, D. J., Norris, R. D., Pierazzo, E., Ravizza, G., Rebolledo-Vieyra, M., Reimold, U., Robin, E., Salge, T., Speijer, R. P., Sweet, A. R., Urrutia-Fucugauchi, J., Vajda, V., Whalen, M. T., and Willumsen, P. S., 2010, Impact and mass extinction: Evidence linking Chicxulub with the Cretaceous-Paleogene boundary: Science, v. 327, p. 1214–1218.

21. McLean, D. M., 1985, Deccan Traps mantle degassing in the terminal Cretaceous marine extinctions: Cretaceous Research, v. 6, p. 235–259;

Courtillot, V., Besse, J., Vandamme, D., Montigny, R., Jaeger, J.-J., and Cappetta, H., 1986, Deccan flood basalts at the Cretaceous/Tertiary boundary?: Earth and Planetary Science Letters, v. 80, p. 361–374; Alvarez, W., 2003, Comparing the evidence relevant to impact and flood basalt at times of major mass extinctions: Astrobiology, v. 3, no. 1, p. 153–161; Gertsch, B., Keller, G., Adatte, T., Garg, R., Prasad, V., Berner, Z., and Fleitmann, D., 2011, Environmental effects of Deccan volcanism across the Cretaceous-Tertiary transition in Meghalaya, India: Earth and Planetary Science Letters, v. 310, no. 3–4, p. 272–285; Richards, M. A., Alvarez, W., Self, S., Karlstrom, L., Renne, P. R., Manga, M., Sprain, C. J., Smit, J., Vanderkluysen, L., and Gibson, S. A., 2015, Triggering of the largest Deccan eruptions by the Chicxulub impact: Geological Society of America Bulletin, v. 127, no. 11/12, p. 1507–1520.

22. Alroy, J., 1999, The fossil record of North American mammals: Evidence for a Paleocene evolutionary radiation: Systematic Biology, v. 48, no. 1, p. 107–118, Fig. 2.

23. Alroy, 1999, op. cit., Fig. 1.

24. For example, Stanley, S. M., 1986, Earth and life through time, New York, W. H. Freeman, 690 p., Fig. 17-6 on p. 532.

25. White, T. D., Asfaw, B., Beyene, Y., Haile-Selassie, Y., Lovejoy, C. O., Suwa, G., and WoldeGabriel, G., 2009, *Ardipithecus ramidus* and the paleobiology of early hominids: Science, v. 326, p. 64, 75–86, with several articles in this issue.

Chapter 8: The Great Journey

1. McNeill, J. R., and McNeill, W. H., 2003, The human web: New York, W. W. Norton, 350 p. See their map on p. 159.

2. Marques, A. P., 1990, Portugal e o descobrimento do Atlântico/Portugal and the discovery of the Atlantic, Lisbon, Imprensa Nacional, Casa da Moeda, 117 p.

3. Leitão, H., and Alvarez, W., 2011, The Portuguese and Spanish voyages of discovery and the early history of geology: Geological Society of America Bulletin, v. 123, p. 1219–1233.

4. Debyser, J., de Charpal, O., and Merabet, O., 1965, Sur le caractère glaciaire de la sédimentation de l'Unité IV au Sahara central: Comptes Rendus Hebdomadaires des Seances de l'Académie des Sciences, v. 261, no. 25, p. 5575–5576.

5. McNeill, J. R., and McNeill, W. H., 2003, op. cit., p. 166.

6. Dreyer, E. L., 2007, Zheng He: China and the oceans in the early Ming Dynasty, 1405–1433, New York, Pearson Longman, 238 p.

7. Bullard, E. C., Everett, J. E., and Smith, A. G., 1965, The fit of the continents around the Atlantic: Royal Society of London Philosophical Transactions, Series A, v. 258, p. 41–51.

8. Anderson, A., Barrett, J. H., and Boyle, K. V., 2010, The global origins and development of seafaring: Cambridge, UK, McDonald Institute for Archaeological Research, 330 p.

9. Klein, R. G., 2009, Darwin and the recent African origin of modern humans: Proceedings of the National Academy of Sciences, v. 106, no. 38, p. 16007–16009.

10. This is the date of the Dmanisi site in the Republic of Georgia: Messager, E., Nomade, S., Voinchet, P., Ferring, R., Mgeladze, A., Guillou, H., and Lordkipanidze, D., 2011, ⁴⁰Ar/ ³⁹Ar dating and phytolith analysis of the early Pleistocene sequence of Kvemo-Orozmani (Republic of Georgia); chronological and palaeoecological implications for the hominin site of Dmanisi: Quaternary Science Reviews, v. 30, no. 21–22, p. 3099–3108.

11. Harmand, S., Lewis, J. E., Feibel, C. S., Lepre, C. J., Prat, S., Lenoble, A., Boës, X., Quinn, R. L., Brenet, M., Arroyo, A., Taylor, N., Clément, S., Daver, G., Brugal, J.-P., Leakey, L., Mortlock, R. A., Wright, J. D., Lokorodi, S., Kirwa, C., Kent, D. V., and Roche, H., 2015, 3.3-million-year-old stone tools from Lomekwi 3, West Turkana, Kenya: Nature, v. 521, p. 310–315.

12. Atkinson, Q. D., 2011, Phonemic diversity supports a serial founder effect model of language expansion from Africa: Science, v. 332.

13. Lordkipanidze, D., Jashashvili, T., Vekua, A., de Leon, M. S. P., Zollikofer, C. P. E., Rightmire, G. P., Pontzer, H., Ferring, R., Oms, O., Tappen, M., Bukhsianidze, M., Agusti, J., Kahlke, R., Kiladze, G., Martinez-Navarro, B., Mouskhelishvili, A., Nioradze, M., and Rook, L., 2007, Postcranial evidence from early *Homo* from Dmanisi, Georgia: Nature, v. 449, no. 7160, p. 305–310.

14. There is some possibility for confusion here, because *H. erectus* was found in China and Java already in the late nineteenth century. When similar fossils were found in Africa in the twentieth century, they were at first called *Homo erectus*, and the evolutionary sequence *H. habilis–H. erectus–H. sapiens* became familiar to those of us who read about human origins a few decades ago. Current usage applies the name *H. ergaster* to the *H. erectus*–like fossils from Africa, and sees *H. erectus* as an Asian descendant of the *H. ergaster* peoples who emigrated from Africa and moved eastward.

15. I learned this interesting result of the field collection of stone tools from Paolo Appignanesi of the Museo Archeologico Statale di Cingoli and Alessandro Montanari of the Osservatorio Geologico di Coldigioco.

16. Beyin, A., 2006, The Bab al Mandab vs the Nile-Levant: An appraisal of the two dispersal routes for early modern humans out of Africa: African Archaeological Review, v. 23, no. 1–2, p. 5–30.

17. Lhote, H., 1959, The search for the Tassili frescoes; the story of the prehistoric rock-paintings of the Sahara, New York, Dutton, 236 p.

18. Fagan, B. M., 2009, Floods, famines and emperors: El Niño and the fate of civilizations, New York, Basic Books, p. 81.

19. For a more detailed explanation of what follows, see Wells, S., 2007, Deep ancestry, Washington, D.C., National Geographic Society, 247 p.

20. Wells, S., 2007, Deep ancestry, Washington D.C., National Geographic Society, p. 40.

21. https://genographic.nationalgeographic.com/human-journey/.

22. Ammerman, A. J., and Cavalli-Sforza, L. L., 1984, The Neolithic transition and the genetics of populations in Europe, Princeton, Princeton University Press.

23. Ammerman, A. J., 2014, Setting our sights on the distant horizon: Eurasian prehistory, v. 11, no. 1–2, p. 203–236.

24. Francis, R. C., 2015, Domesticated: Evolution in a man-made world, New York, W. W. Norton, 484 p., Ch. 11.

25. Anthony, D. W., 2007, The horse, the wheel, and language, Princeton, Princeton University Press, 553 p., Ch. 10.

26. Outram, A. K., Stear, N. A., Bendrey, R., Olsen, S., Kasparov, A., Zaibert, V., Thorpe, N., and Evershed, R. P., 2009, The earliest horse harnessing and milking: Science, v. 323, p. 1332-1335.

27. Crosby, A. W., 1972, The Columbian Exchange: Biological and cultural consequences of 1492, Westport, Conn., Greenwood Pub. Co., 268 p.

Chapter 9: Being Human

1. Deacon, T. W., 1997, The symbolic species: The co-evolution of language and the brain, New York, W. W. Norton, p. 23.

2. The only textbook of Big History that has been published so far is built around the concept of thresholds: Christian, D., Brown, C. S., and Benjamin, C., 2014, Big History: Between nothing and everything, New York, McGraw Hill, 332 p.

3. Christian, D., 2004, Maps of time. An introduction to Big History, Berkeley, University of California Press, 642 p.

4. Christensen, M. H., and Kirby, S., 2003, Language evolution: The hardest problem in science?, *in* Christensen, M. H., and Kirby, S., eds., Language evolution: Oxford, Oxford University Press, p. 1–15.

5. These criteria come from Janson, T., 2012, The history of languages: an introduction, Oxford, Oxford University Press, 280 p. However, Janson uses 2 million years ago as the age of the earliest stone tools, and recent discoveries have pushed the earliest tools back to 3.3 million years ago: Harmand, S., Lewis, J. E., Feibel, C. S., Lepre, C. J., Prat, S., Lenoble, A., Boes, X., Quinn, R. L., Brenet, M., Arroyo, A., Taylor, N., Clement, S., Daver, G., Brugal, J.-P., Leakey, L., Mortlock, R. A., Wright, J. D., Lokorodi, S., Kirwa, C., Kent, D. V., and Roche, H., 2015, 3.3-million-year-old stone tools from Lomekwi 3, West Turkana, Kenya: Nature, v. 521, no. 7552, p. 310–315.

6. Wade, N., 2006, Before the dawn, London, Penguin, Ch. 10; Anthony, D. W., 2007, The horse, the wheel, and language, Princeton, Princeton University Press, 553 p.

7. Bryson, B., 1990, The mother tongue: English and how it got that way, New York, Avon, 270 p.; Ostler, N., 2005, Empires of the word: A language history of the world, New York, HarperCollins, 615 p.; Janson, T., 2012, The history of languages: An introduction, Oxford, Oxford University Press, 280 p.; Nadeau, J.-B., and Barlow, J., 2013, The story of Spanish, New York, Saint Martin's Griffin, 418 p.

8. Goudsblom, J., 1992, Fire and civilization: London, Penguin, 247 p.

9. Pyne, S. J., 2001, Fire: A brief history, Seattle, University of Washington Press, 204 p.; Bowman, D. M. J. S., Balch, J. K., Artaxo, P., Bond, W. J., Carlson, J. M., Cochrane, M. A., D'Antonio, C. M., DeFries, R. S., Doyle, J. C., Harrison, S. P., Johnston, F. H., Keeley, J. E., Krawchuk, M. A., Kull, C. A., Marston, J. B., Moritz, M. A., Prentice, I. C., Roos, C. I., Scott, A. C., Swetnam, T. W., van der Werf, G. R., and Pyne, S. J., 2009, Fire in the Earth system: Science, v. 324, p. 481–484.

10. Canfield, D. E., 2014, Oxygen: A four billion year history, Princeton, Princeton University Press, 196 p.

11. Abraham, D. S., 2015, The elements of power: Gadgets, guns and the struggle for a sustainable future in the rare metals age, New Haven, Yale University Press, 336 p.

12. For a Big History view of the Industrial Revolution, see Christian, D., Brown, C. S., and Benjamin, C., 2014, Big History: Between nothing and everything, New York, McGraw Hill, Ch. 11.

13. Carlos Camargo pointed this out to me: http://www.iceman.it/en.

14. Homer, 1946, The Odyssey, translated by E. V. Rieu, Baltimore, Penguin, p. 26.

15. Nur, A., 2008, Apocalypse: Earthquakes, archaeology, and the wrath of God: Princeton, Princeton University Press, 309 p.; Drews, R., 1993, The end of the Bronze Age: Changes in warfare and the catastrophe ca. 1200 B.C.: Princeton, Princeton University Press, 252 p.

16. For a remarkable compilation of many features of Earth history, including useful resource deposits, see Bradley, D. C., 2011, Secular trends in the geologic record and the supercontinent cycle: Earth-Science Reviews, v. 108, p. 16–33.

17. Constantinou, G., 1982, Geological features and ancient exploitation of the cupriferous sulphide orebodies of Cyprus, *in* Muhly, J. D., Maddin, R., and Karageorghis, V., eds., Early metallurgy in Cyprus, 4000–500 B.C.: Larnaca, Cyprus, Pierides Foundation, p. 13–24.

18. Moran, W. L., 1992, The Amarna letters, Baltimore, Johns Hopkins University Press, 393 p., letter ESA 35, p. 107.

19. Wilson, R. A. M., 1959, The geology of the Xeros-Troodos area: Geological Survey Department Cyprus Memoir, v. 1, p. 1–135.

20. Hess, H. H., 1960, The evolution of ocean basins (preprint): Princeton University, Department of Geology; Hess, H. H., 1962, History of ocean basins, *in* Engel, A. E. J., James, H. L., and Leonard, B. F., eds., Petrologic studies: A volume in honor of A. F. Buddington: Geological Society of America, p. 599–620.

21. Gass, I. G., 1968, Is the Troodos massif of Cyprus a fragment of Mesozoic ocean floor?: Nature, v. 220, p. 39–42; Moores, E. M., and Vine, F. J., 1971, The Troodos Massif, Cyprus and other ophiolites as oceanic crust: Evaluation and implications: Philosophical Transactions of the Royal Society of London, series A, v. 268, p. 443–466.

22. Ballard, R. D., and Grassle, J. F., 1979, Incredible world of the deep-sea rifts: National Geographic, v. 156, no. 5 (November), p. 680–705, picture on p. 702–703.

23. Krasnov, S. G., Cherkashev, G. A., Stepanova, T. V., Batuyev, B. N., Krotov, A. G., Malin, B. V., Maslov, M. N., Markov, V. F., Poroshina, I. M., Samovarov, M. S., Ashadze, A. M., Lazareva, L. I., and Ermolayev, I. K., 1995, Detailed geological studies of hydrothermal fields in the North Atlantic, *in* Parson, L. M., Walker, C. L., and Dixon, D. R., eds., Hydrothermal vents and processes: London, Geological Society of London Special Publication no. 87, p. 43–64.

24. Little, C. T. S., Cann, J. R., Herrington, R. J., and Morisseau, M., 1999, Late Cretaceous hydrothermal vent communities from the Troodos Ophiolite, Cyprus: Geology (Boulder), v. 27, no. 11, p. 1027–1030.

25. Yener, K. A., 2000, The domestication of metals: The rise of complex metal industries in Anatolia, Leiden, Boston, Brill, 210 p.

Chapter 10: What Was the Chance of All This Happening?

1. A well-known and very elaborate attempt to formulate laws governing human history (though not mathematical ones) was carried out by Arnold Toynbee (1934–1961, A study of history, London, Oxford University Press). Graeme Snooks has reviewed the current status of this quest (1998, The laws of history, London, Routledge, 308 p.).

2. Gould, S. J., 1987, Time's arrow, time's cycle, Cambridge, Mass., Harvard University Press, 222 p.

3. Many such data sets have been compiled by Bradley, D. C., 2011, Secular trends in the geologic record and the supercontinent cycle: Earth-Science Reviews, v. 108, p. 16–33, and in its online supplement.

4. Muller, R. A., and MacDonald, G. J., 2000, Ice ages and astronomical causes, New York, Springer, figures in Ch. 1.

5. The English victory over the Spanish Armada began with a fireship attack at Calais on the night of August 7–8, 1588, made possible by the way the wind was blowing. Marching orders from General Robert E. Lee were lost

and then recovered by Union soldiers in September 1862, giving the Union an advantage at the subsequent Battle of Antietam.

6. There is a clear explanation of power-law distributions as applied to wars in Pinker, S., 2011, The better angels of our nature: Why violence has declined, New York, Viking, p. 210–215.

7. Large, significant impacts are rare because there are far more small objects orbiting the Sun than large ones. Sand-size impactors are very abundant, but mountain-size ones are extremely rare. The B612 Foundation, started by former astronaut Ed Lu, aims to find all dangerous potential impactors and, if one is found that threatens Earth, to mount a space mission to divert it. The foundation's website is http://b612foundation.org/. The present-day size distribution of potential impactors is the result of its own complicated history. All size ranges were more abundant in the early solar system, before most of the debris was removed by hitting something else or being flung out of the solar system. This history of cleaning up explains the crater-saturated surface of the Moon, where little has happened since it was very young, making the Moon's surface a museum of early solar system history. Early Earth must have looked like that, too, before our planet's intense geological history wiped out almost every trace of its remote past. But there is a further complication: We now know that at least once, in the Ordovician, about 466 million years ago, there was a collision in the asteroid belt that shattered the two objects that collided and greatly increased the number of asteroid fragments in the inner solar system. This led to an unusual number of cratering events on Earth and to both meteorites and meteorite debris in sediments of that age, first discovered in Sweden by Mario Tassinari and Birger Schmitz (Schmitz, B., and Tassinari, M., 2001, Fossil meteorites, *in* Peucker-Ehrenbrink, B., and Schmitz, B., eds., Accretion of extraterrestrial matter throughout Earth's history: New York, Kluwer, p. 319–331). Birger Schmitz is now studying the abundance and chemistry of extraterrestrial spinel grains in the sedimentary record of the last half billion years, aiming to reconstruct the history of impacts on Earth—both their frequency and the types of impactors, which he can recognize from the chemistry of the meteoritic spinel grains in the sediment (Schmitz, B., 2013, Extraterrestrial spinels and the astronomical perspective on Earth's geological record and evolution of life: Chemie der Erde, v. 73, p. 117–145).

8. Prigogine, I., and Stengers, I., 1984, Order out of chaos: Man's new dialogue with nature, Toronto, Bantam, section II.4, "Laplace's Demon," p. 75–77.

9. Lorenz, E., 1993, The essence of chaos, Seattle, University of Washington Press, 227 p.; Briggs, J., and Peat, F. D., 1989, Turbulent mirror: An illustrated guide to chaos theory and the science of wholeness, New York, Harper and Row, 222 p.

10. Zhang, D., Györgyi, L., and Peltier, W. R., 1993, Deterministic chaos in the Belousov-Zhabotinsky reaction: Experiments and simulations: Chaos, v. 3, no. 4, p. 723–745. Many videos of the evolving patterns of the Belousov-Zhabotinsky reaction are available on the Web.

11. May, R. M., 1976, Simple mathematical models with very complicated dynamics: Nature, v. 261, no. 5560, p. 459–467. For a clear explanation of the behavior of the iterated logistic equation, see Gleick, J., 1987, Chaos: Making a new science, New York, Viking, p. 69–77.

12. Technically, since today's birds are descended from dinosaurs, it is actually only the nonavian dinosaurs that became extinct 66 million years ago: Dingus, L., and Rowe, T., 1998, The mistaken extinction: Dinosaur evolution and the origin of birds, New York.

13. Alvarez, L. W., Alvarez, W., Asaro, F., and Michel, H. V., 1980, Extraterrestrial cause for the Cretaceous-Tertiary extinction: Experimental results and theoretical interpretation: Science, v. 208, p. 1095–1108; Smit, J., and Hertogen, J., 1980, An extraterrestrial event at the Cretaceous-Tertiary boundary: Nature, v. 285, p. 198–200; Alvarez, W., 1997, T. rex and the Crater of Doom, Princeton, Princeton University Press, 185 p.; Schulte, P., et al., 2010, Impact and mass extinction: Evidence linking Chicxulub with the Cretaceous-Paleogene boundary: Science, v. 327, p. 1214–1218. It looks increasingly like the huge Deccan volcanic outpourings in India that were going on at the time of the impact and extinction may have been affected, though not caused, by the impact and may possibly have contributed to the extinction: Richards, M. A., Alvarez, W., Self, S., Karlstrom, L., Renne, P. R., Manga, M., Sprain, C. J., Smit, J., Vanderkluysen, L., and Gibson, S. A., 2015, Triggering of the largest Deccan eruptions by the Chicxulub impact: Geological Society of America Bulletin, v. 127, no. 11/12, p. 1507–1520.

14. Perihelion distances of 680,062 known asteroids and 3,330 visible comets are from NASA's JPL Small-Body Database Search Engine (http://ssd.jpl.nasa.gov/sbdb_query.cgi#x).

15. Gleick, J., 1987, Chaos, New York, Viking, 352 p.; Lorenz, E., 1993, The essence of chaos, Seattle, University of Washington Press, 227 p.

16. Shubin, N., 2008, Your inner fish: A journey into the 3.5-billion-year history of the human body, New York, Random House, 240 p.

17. Calvin, W. H., 1987, The brain as a Darwin machine: Nature, v. 330, p. 33–34.

18. Plotkin, H., 1994, Darwin machines and the nature of knowledge, Cambridge, Mass., Harvard University Press, 269 p.

19. Roep, T. B., Holst, H., Vissers, R. L. M., Pagnier, H., and Postma, D., 1975, Deposits of southward-flowing Pleistocene rivers in the channel region, near Wissant, NW France: Palaeogeography, Palaeoclimatology, Palaeoecology, v. 17, no. 4, p. 289–308; Smith, A. J., 1985, A catastrophic origin for the palaeovalley system of the eastern English Channel: Marine Geology, v. 64, no. 1–2, p. 65–75.

20. Gupta, S., Collier, J. S., Palmer-Felgate, A., and Potter, G., 2007, Catastrophic flooding origin of shelf valley systems in the English Channel: Nature, v. 448, no. 7151, p. 342–345. A nontechnical account is available online.

21. Bryson, B., 2003, A short history of nearly everything, New York, Broadway Books, p. 1–4.

22. I've used conservative estimates: About 10^9 women will contribute to the next generation, each having about 10^3 eggs, for a total of 10^{12} eggs. Each woman might meet something like 10^3 men, and there are about 10^8 sperm trying to get there first when conception occurs, on each of perhaps 10^2 possible occasions, for a total of 10^{13} sperm. Multiplying the number of eggs times the number of sperm gives about 10^{25} possible combinations.

23. By coincidence, this number, 10^{100}, has been called a "googol" and offered as an example of an extremely large number: Kasner, E., and Newman, J. R., 1940, Mathematics and the imagination, New York, Simon and Schuster, p. 20–25. Carl Sagan has also discussed the googol: Sagan, C., 1980, Cosmos, New York, Random House, p. 219–220.

Index

Page numbers in *italics* refer to illustrations.